STATISTICS

WITH MICROSOFT® EXCEL

THIRD EDITION

BEVERLY J. DRETZKE

PEARSON
Prentice Hall

Upper Saddle River, NJ 07458

Supplement Editor: Joanne Wendelken
Executive Editor: Petra Recter
Editor-in-Chief, Math: Sally Yagan
Vice President of Production & Manufacturing: David W. Riccardi
Executive Managing Editor: Kathleen Schiaparelli
Managing Editor: Nicole Jackson
Production Editor: Zachary Hubert
Supplement Cover Manager: Paul Gourhan
Supplement Cover Designer: Joanne Alexandris
Manufacturing Buyer: Ilene Kahn

© 2005 Pearson Education, Inc.
Pearson Prentice Hall
Pearson Education, Inc.
Upper Saddle River, NJ 07458

Printed in the United States of America

10 9 8 7 6 5 4 3 2

ISBN 0-13-147111-2

Pearson Education Ltd., *London*
Pearson Education Australia Pty. Ltd., *Sydney*
Pearson Education Singapore, Pte. Ltd.
Pearson Education North Asia Ltd., *Hong Kong*
Pearson Education Canada, Inc., *Toronto*
Pearson Educación de Mexico, S.A. de C.V.
Pearson Education—Japan, *Tokyo*
Pearson Education Malaysia, Pte. Ltd.

Preface

The third edition of *Statistics with Microsoft® Excel* was prepared using the Office 2003 version of Microsoft® Excel. The main differences between Excel 2003 and earlier versions will be noticed when using Pivot Table, Pivot Chart, and functions. I have included many screen prints, however, so users should have no difficulties following the instructions.

I have used *Statistics with Microsoft® Excel* in my introductory statistics classes for several years, and some of the revisions you see in the third edition were made in response to students' comments. For example, Chapter 3 includes a section entitled Using Formulas in Statistics. In this section, I explain how to use formulas for computations that students typically carry out using a hand-held calculator—deviation scores, squared deviation scores, Z scores. Only a few students realized that the completed Excel worksheet could function as a template for different sets of data. To help students see that they could use the same worksheet over and over again, I specifically use the term *template* and illustrate its use with data sets of different sizes.

In Chapter 4, I have included a section on using the Histogram Tool to create a frequency distribution and histogram for ungrouped data as well as grouped data. In Chapter 5, I explain how to substitute population values for sample values in the Descriptive Statistics Tool output. I added a section on curvilinear regression in Chapter 11. Chapter 13 is a new chapter that explains how to generate random numbers.

My former students tell me that they often use Excel to produce tables and graphs for their research papers. Through experience, I have learned that the user can modify many aspects of tables and charts—titles, font size, legends, item order, trendlines, and so on. I have expanded the sections on tables and graphs to introduce the users to many of these enhancements.

Over the years, I have received e-mail messages from many users of *Statistics with Microsoft® Excel*. Several times, I have been asked to explain how to recode a quantitative variable into a categorical variable. Recoding is really easy with Excel's LOOKUP function, and I have included step-by-step instructions on how to do it.

I am very grateful to my students for their comments and suggestions. I am also grateful to other users who have contacted me with their questions and suggestions. I also wish to thank Petra Recter and Joanne Wendelken of Prentice-Hall for their assistance.

If you have any comments or suggestions that you would like to make regarding this manual, please send me a message.

Beverly J. Dretzke

(dretzkbj@uwec.edu)

Contents

Chapter 1: Getting Started

Chapter 2: Entering, Editing, and Recoding Information

Chapter 3: Formulas

Chapter 4: Frequency Distributions

Chapter 5: Descriptive Statistics

Chapter 6: Probability Distributions

Chapter 7: Testing Hypotheses About One Sample Means

Chapter 8: Testing Hypotheses About the Difference Between Two Means

Chapter 9: Analysis of Variance

Chapter 10: Correlation

Chapter 11: Regression

Chapter 12: Cross Tabulations

Chapter 13: Random Samples

Chapter 1

Getting Started

Introduction and Overview

This manual describes how to use Excel to perform common statistical procedures. Detailed, step-by-step examples are provided to illustrate how to use formulas, functions, the Pivot Table, the Pivot Chart, and Data Analysis Tools for statistical applications. This manual is designed to be used along with a statistics textbook, and, therefore, provides only brief explanations of the analyses themselves.

This manual contains two major parts: 1) a general introduction to Excel that I hope will provide enough information for you to be able to use the Excel worksheet and 2) chapters that describe how to use Excel to carry out statistical analysis procedures.

The first three chapters introduce worksheets, and describe how to enter data, use simple formulas to manipulate data, save, retrieve, and print. Chapter 4 shows how to organize data using frequency distributions and frequency graphs. Chapter 5 presents descriptive statistics. Chapter 6 discusses probability distributions and the ways that Excel can provide information about them. Chapter 7 presents techniques for testing hypotheses about one-sample means, followed by Chapter 8, which explains how to use Excel to test hypotheses about the difference between two means. Analysis of variance for simple cases is covered in Chapter 9. The Pearson correlation coefficient and the Spearman rank correlation coefficient are presented in Chapter 10. Chapter 11 includes bivariate regression, multiple regression, curvilinear regression, and dummy coding of qualitative variables. Cross tabulations and the chi-square test of independence are discussed in Chapter 12. Finally, Chapter 13 presents random sampling procedures.

Versions of Excel

The Microsoft Office 2003 version of Excel was used when preparing this manual and all of the screens pictured in this manual are from that version. If you are using an earlier version, you will note some slight differences, especially in the Pivot Table and the function dialog boxes.

Versions of Windows

The copies of screens shown in this book were taken from a PC using Windows 2000. The screens will appear slightly different if you are using an earlier Windows version or a Mac. After you are operating comfortably within Excel, these differences should be minor. There will, however, be slight differences between Macs and PCs in the keys used for commands.

What May Be Skipped

If you have used spreadsheets before, you can probably omit much of the first three chapters. Other programs, such as Lotus 1-2-3 and Quattro, use somewhat different terminology to describe the operations covered in Chapters 2 and 3. The concepts, however, are essentially the same. For those of you who are looking for a specific procedure to use for a statistical test, each of the chapters can be used independently.

Excel Worksheet Basics

Using the Mouse

Mice come in several forms. The majority are provided with new computers and roll on a desktop or pad. When rolled, a small ball on the bottom causes the pointer on the screen (called the screen pointer) to move in a corresponding way. Another version (called a trackball) places a larger ball in a framework that allows you to roll the ball with your fingers. Other forms have small screens that you move your finger across as you move the mouse. The pressure of your finger moving across the screen causes a screen pointer to move in synchrony with your movements.

All devices have at least one button (and most likely two or more) that you can click or hold down, sometimes while also moving the ball. Five basic actions operate the mouse.

- **Point**. You point to objects on the screen by sliding the mouse on the deskpad or by rolling the trackball. As the screen pointer tracks the movements made on your desk, its shape changes. Most often the shape will be the outline of an arrow or the outline of a plus sign. The shape usually changes when a task is completed.

- **Click**. "Click" means to press and release the left mouse button (called a left-click). If you are pointing at an executable command, this action causes it to take place. If you point to any cell on the worksheet and click, that cell becomes the active cell and is ready to receive data.

- **Double-click**. "Double-click" means to press and release the left mouse button twice rapidly. If you fail to press rapidly enough, it is interpreted as one click. The double-click often replaces the two-step sequence of selecting a command and then clicking on OK to execute that command.

- **Right-click**. "Right-click" means to press down on the right button of the mouse. A right-click is often used to display special shortcut menus.

- **Drag**. Objects on the screen are moved by dragging. To drag, place the mouse pointer on the item you want to move, click and hold down the left mouse button—do NOT release it. While you hold the mouse button down, slide the mouse to move the screen pointer and the item to the location you want. Then release the mouse button.

Starting Excel

To start the program using Windows 2000, click on **Start** in the lower left of the screen. Move the mouse to **Programs** and then continue moving through menus until you find the **Microsoft Excel** icon to click and begin.

Exiting Excel

There are two ways that you can exit Excel.

1. Click the upper **X** (the **close button**) that you find in the upper right corner of the screen. If two sets of boxes are showing, the lower set applies to the worksheet that is showing while the upper one is for the application or program itself (i.e., Excel). If you have edited (changed) any of the information in the workbook, you will be prompted to save the information before closing the program.

2. Click **File** in the Menu Bar and select **Close** and then **File** and select **Exit**. You will be prompted to save the file if you edited any information in the workbook.

Layout of Worksheets and Workbooks

The figure shown on the next page displays what is on the computer screen when Excel 2003 is opened. Due to differences in computers, the display may be somewhat different on your screen. A brief description of each of the items follows the figure. We'll start at the top of the screen and work our way down to the bottom.

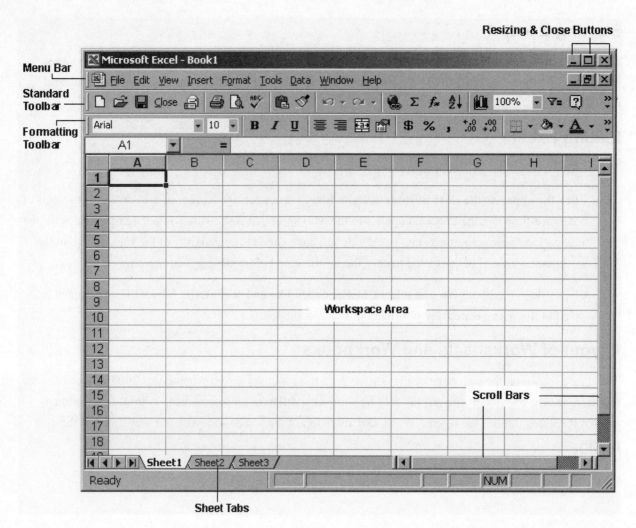

- **Program Title Bar**. This is most likely at the very top of the screen. The default title is Microsoft Excel - Book 1. It indicates the name of the application and that you are in what is called Book 1, the name given to a newly opened worksheet. Each book initially consists of three worksheets (this can be changed) that are stored together as a unit called a book. When you save your work, all of the sheets in this book will be saved together as one file.

- **Resizing and close buttons**. The three icons are at the far right in the Program Title Bar and at the far right in the Menu Bar. A description is provided of each icon.

  - The minimize button "shrinks" the application, which is then represented as a button on the taskbar at the bottom of the screen. The application is inactive but not closed. It can be quickly reactivated by clicking on the button.

  - The maximize button switches (toggles) between the full-screen and window views.

  - The close button closes the Excel application. You will be asked if you want to save your work, if there is any, before the application is closed.

- **Main Menu Bar**. Commands are grouped into categories such as **File, Edit, View**, etc. Clicking on one of these will drop down the commands in that group.

- **Standard Toolbar**. This is a ribbon of icons that are designed to ease your access to commands. Some are not as obvious as one might wish, but all you need to do to find out what the button does is to point at it with your mouse pointer and pause. Its name is then displayed. For example, the first icon, labeled New, resembles a sheet of paper and will create a new workbook if it is clicked. The second will open a file, etc. **Microsoft Excel Help** (looks like a question mark), near the extreme right of the Toolbar, will give you information about how to do something. You will be able to obtain assistance by accessing Contents, Answer Wizard, or Index.

- **Formatting Toolbar**. Commands that change the appearance of text are found in the Formatting Toolbar. The default font is displayed (initially Arial in most cases). If you wish to use a different font, you need only click on the down arrow next to the font window to see which ones are available and select one. To change the default font for the whole worksheet, use Help and search for the term Fonts.

  The size of the font (10) is displayed in the next window with a down arrow for changing it.

  Following this are buttons for bold, *italic*, <u>underline</u>, left-alignment, center-alignment, right-alignment, and centering across columns. The next five icons control the formatting of numbers—currency, percentage, use of commas, and the number of decimals displayed. To place any of a variety of borders around a cell or cells, use the next icon. The last two icons deal with colors, which we will not cover.

- **Formula Bar**. The first window in the Formula Bar shows the address of the active cell. Initially it is A1, so this is the address you see displayed. If you move to a different cell and click in it to make it the active cell, then that cell's address will be displayed. The first box in the formula bar is also called the **Name Box**. Notice what happens to the Formula Bar when you type some numbers in cell A1. Three more buttons appear.

  - The **red X** is clicked when you want to delete the information you have typed in the active cell. That information is also displayed in the Formula Bar and will be removed from both sites if you click on the X.

  - The **green check mark**, when clicked, indicates that the data as entered are acceptable. The data will remain in the cell, but the three buttons disappear, indicating that the cell is not being edited.

  - The **equal sign** enables you to edit a formula that has been entered in the active cell.

- **Worksheet Area**. The worksheet consists of 16,777,216 cells with columns labeled as letters and rows as numbers. There are 256 columns and 65,536 rows. Each cell is identified by the combination of its column letter and row number.

- **Scroll Bars** are found at the right and at the bottom of the worksheet. There are two small arrows (which look like a triangle on its side) at either end of each scroll bar that, when clicked with the mouse, move the screen up, down, right, or left one line for each click. Some users call the box within the scroll bar an elevator. You can grab the box (elevator) and drag it. The screen will move a distance that corresponds to the amount you move the box.

- **Worksheet Tabs**. These are at the lower left of the screen and are labeled as Sheet 1, Sheet 2, etc. If you insert several additional worksheets, you can move through them by clicking the arrows to the left of the sheet tabs. Two arrows move the active sheet to the left: One moves a sheet at a time and the other will move to the left-most sheet. The same applies to the right arrows. All the worksheets are stored together as one unit, called a Book.

- **Status Bar**. The Status Bar indicates what is happening. Most of the time, when nothing special is being done, it simply says **Ready** at the left of the screen. The word **Enter** will be displayed when you are typing a cell entry.

Dialog Boxes

Dialog Boxes usually require that you choose from a number of alternatives or enter your choices. Many of the statistical analysis procedures that are presented in this manual are associated with commands that are followed by dialog boxes. For example, if you click **Insert** and select **Function**, a dialog box like the one shown below will appear. You will usually select a function category and a function name. You make your selections by clicking on them.

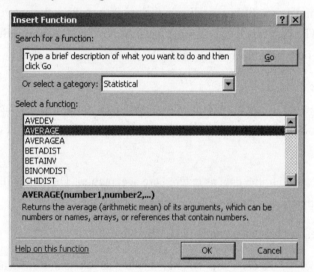

When you click the OK button at the bottom of the dialog box, another dialog box will be displayed that asks you to provide information regarding location of the data in the Excel worksheet.

Loading Excel's Analysis Toolpak

The Analysis Toolpak is an Excel Add-In that may not necessarily be loaded on your computer. If **Data Analysis** does not appear in the Tools menu, then click on **Add-Ins** in the Tools menu to get the Add-Ins dialog box shown below.

Click in the box to the left of **Analysis ToolPak** to place a checkmark there. Then click **OK**. The ToolPak will load and will be listed at the bottom of the Tools menu as **Data Analysis**.

Chapter 2

Entering, Editing, and Recoding Information

Opening Documents

Opening a Brand-New Worksheet

If you started Excel by clicking on the icon, the screen opened with a new, blank worksheet. The title at the top indicates it is called **Book 1**. Along the bottom are **sheet tabs** labeled **Sheet 1**, **Sheet 2**, etc. The sheets are stored together in one unit, called a book. You might, for example, choose to keep all data analysis for one research project together in one unit, now called **Book 1**, but renamed by you as **Opinion Survey**. You could create other books as well that contain the data and analyses for other research projects, records of project expenses, course assignments, and so on.

If you click on the X in the upper right of the Excel window, indicating that you want to quit, the program will ask you if you want to save the file you have created. If you have already saved it, and therefore have named it, Excel will save it using the same name unless you use the **Save As** command, which is used to change the name or location for saving. The Save As option is located in the File menu.

Opening a File You Have Already Created

If you want to continue working on a file that you previously created, you will click **File,** select **Open**, then locate and select the file. You can also go to the file by clicking on the Open button, which looks like this:
.

Entering Information

Addresses

Columns are identified by letters of the alphabet that are shown across the top of the worksheet. Two letter combinations are used (e.g., AA, AB) after Z is reached. This continues until the final column is reached, which is the 256th column and is labeled IV. Rows continue numerically until the last row (65,536) is reached. Each cell is identified by a combination of a letter and a number. For example, cell A1 is located in column A, row 1, and cell D15 is located in column D, row 15.

Activating a Cell or Range of Cells

When the worksheet is initially opened, cell A1 is automatically the **active cell**. It has a dark outline around it, which indicates that whatever you type will be entered into that cell. The address A1 is displayed in the **Name Box,** which is to the left of the **Formula Bar**.

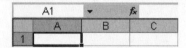

Move to cell C2, click, and note that C2 is displayed in the Name Box shown on the right.

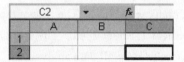

We often need to refer to more than one cell at a time. A group of cells is called a **Range**. Click on cell B4, hold the mouse button down, and drag to B8. Release the button. The worksheet is now **marked** (highlighted) in the range of cells as shown below. Note that the top cell of the range is not darkened. To indicate the addresses of a range of cells, we separate the address of the upper left and the lower right cell with a colon. Here we have a range for a single column identified as **B4:B8**, although the Name Box displays only the address of the top cell (B4).

To activate cells in many rows and columns (i.e., a range), place the cursor in the upper left cell and drag to the lower right. Now all of the cells in that range will appear highlighted. You can click and drag in the opposite direction if you wish; that is, you can drag from the bottom to the top. This is a good way to proceed if you tend to overshoot your target and move past where you want to end, as many of us do. Another way to activate a range of cells is to click in the upper left cell, press and hold down the **SHIFT** key, and then move to the lower right cell using the scroll bars or arrow keys.

Types of Information

There are three types of information that may be entered into a cell:

1. **Text**. This term refers to alphabet characters (e.g., Exam) or a combination of alphabet, numeric, and symbol characters (e.g., Exam #2).

2. **Numeric**. Any cell entry comprised completely of numbers falls into the numeric category.

3. **Formulas**. These create new information generated from operations performed on numbers that are entered directly in the formula or that are entered in cells of the worksheet.

Filling Adjacent Cells

Filling can mean one of two things. First, filling can mean that you take the contents of a given cell or range of cells and make copies of that material in adjacent cells. Second, filling can also mean that you continue a series or sequence into adjacent cells.

Filling the Same Content into Adjacent Cells

If you have three different treatment conditions and you wish to repeat them in the first column of a worksheet, you would proceed as follows.

1. In a new worksheet, enter **Drug A** in cell A1, **Drug B** in cell A2, and **Placebo** in cell A3.

3. Click in cell A1 and drag to cell A3, to activate that range.

4. Use the **fill handle** to copy the contents of A1:A3 to cells A4 through A12. To do this, move the mouse pointer to the lower right corner of cell A3. It will turn into a black plus sign. Press the left mouse key and hold it down while you drag to cell A12. Release the mouse key and the names of the three conditions will be repeated in column A.

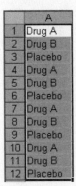

Filling a Series into Adjacent Cells

The subject numbers assigned to research participants are often in numerical order, such that the first subject is 1, the second is 2, and so on.

1. Enter **1** in cell A1 and **2** in cell A2.

	A
1	1
2	2

2. Mark cells A1 and A2 by dragging over them.

	A
1	1
2	2

3. Use the fill handle to continue the series in cells A3 through A10.

	A
1	1
2	2
3	3
4	4
5	5
6	6
7	7
8	8
9	9
10	10

Series

The previous activities show how easily you can repeat cell content in adjacent cells or fill adjacent cells with a series of numbers. The **Series** command can also be used generate a series of numbers. Follow these steps to practice using this command.

1. Enter **5**, **7**, and **9** in cells A1, A2, and A3, respectively.

	A
1	5
2	7
3	9

2. Activate the range **A1:A10**.

	A
1	5
2	7
3	9
4	
5	
6	
7	
8	
9	
10	

3. Click **Edit** and select **Fill→Series**. The Series dialog box will appear. Excel has correctly detected that the series is in Columns, that it is a Linear series, and that it is changing by a step value of 2. If necessary, these entries can be revised.

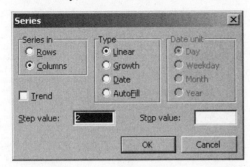

4. Click **OK**. The marked range in the worksheet now contains the series of odd numbers beginning with 5 and ending with 23.

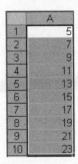

Editing Information

Changing Information

To change information in a cell, you have to consider which situation exists.

- If you have not yet "accepted" the information by clicking the green check mark (or pressing [Enter], or using arrow keys, or …), then you can simply use the backspace or delete key to remove entries.

- If you are typing information into a cell and decide you want to start over, click on the red X (to the left of the Editing Bar) and everything will be deleted.

- If you want to delete everything in the active cell or range of cells, press the [Delete] key.

- If you have already entered and accepted data in a cell, but now want to edit it but not erase all of it, first activate the cell. Then, in the Editing Bar, move the mouse pointer to where you want to make the changes, and click to insert the I-beam.

As a practice exercise for changing information, try going through the steps presented below.

1. In cell A1 type **12346**. Press [**Enter**], which moves you to cell A2.

2. Assume you really wanted to enter 123456. Return to **A1** by using the arrow key or mouse pointer.

3. In the Editing Bar, move the mouse pointer so the I-beam is between 4 and 6. Click once. Note that three editing keys are now shown to the left of 12346.

A1	▼ ✕ ✔ ƒx 12346	
	A	B	C
1	12346		
2			

4. Type **5**, which will be inserted between 4 and 6.

	▼ ✕ ✔ ƒx 123456	
	A	B	C
1	123456		
2			

5. Press [**Enter**].

Moving and Copying Information

A basic principle is that you first indicate which material will have something done to it by marking it, and then you execute the command that does something to the highlighted material. You will see this principle operate in several other places in Excel. When you want to move information, you can do it so that it is removed from one location and placed in another. This is a **cut**. If you want to make a copy of the material so that it is in the original location as well as in another location, you make a **copy**.

1. To copy the entry in cell A1 (123456) to cell B1, first mark cell A1 by clicking on it.

	A	B
1	123456	
2		

2. Move the mouse pointer to the copy icon, 🖻 , or click **Edit** in the Menu Bar and select **Copy**. The outline around the cell will become like the lights on a marquee—it will alternately darken and lighten. It is ready to be copied.

	A	B
1	123456	
2		

3. Click the cell where you want the copy placed, cell **B1**.

	A	B	C
1	123456		
2			

4. Click **Edit** in the Menu Bar and select **Paste** or use the paste icon, 🖻 , to place a copy there.

	A	B	C
1	123456	123456	
2			

5. The copy command operates like a rubber stamp; a copy is stored on the Clipboard and you can continue to place copies anywhere you wish. Activate another cell, say, C3, and paste another copy there.

Cutting is done in a similar fashion, except that the cell is empty after you cut the data from it.

1. Click on a cell with content. For this example, click on cell **A1**.

	A	B	C
1	123456	123456	
2			

2. Click **Edit** in the Menu Bar and select **Cut**. The cell will be outlined, as it was when you used Copy.

	A	B	C
1	123456	123456	
2			

3. Activate the cell where you would like the information to be placed. For this example, click on cell **A3**.

	A	B	C
1	123456	123456	
2			
3			
4			

4. Click **Edit** in the Menu Bar and select **Paste** or click on the paste icon.

	A	B	C
1		123456	
2			
3	123456		
4			

If you cut or copy a group of cells (i.e., a range), the principle is the same. Instead of activating one cell, you click and drag over the cells so that a range of cells is marked. Pasting a range is the same, except that you specify only the upper left cell of the target location.

Moving material between worksheets is accomplished the same way: Mark, indicate Cut or Copy, move to the new sheet and the desired location for the upper left cell, and then paste.

Dragging and Dropping

If you want to move or copy a range of cells on the same sheet, a shortcut is called **drag and drop**. To move a range, mark the cells and position the cursor near an edge, where it will become an outline arrow. Hold down the left key of the mouse and drag to the new location. Release the mouse key. To copy, hold down the **CTRL** key at the same time. If you attempt to drag and drop material into a spot that currently has data, you will get a message that asks if you want to do this, because the old material will be removed. If you have mistakenly done this, you can fall back on the **Undo Drag and Drop** command that you will find in the **Edit** menu.

Inserting or Deleting Rows and Columns

Inserting or deleting rows and columns is relatively easy. If you have entered data in a row that includes columns A, B, and C and want to place a new column between A and B, do the following. Click on the letter, **B** to mark that column. In the Menu Bar, click **Insert** and select **Columns**. A column will be inserted to the left of the marked column. If you want to insert two columns between A and B, simply mark columns B and C at the same time and follow the procedure described above. To delete these two empty columns, mark them and go to the **Edit** command and click **Delete**.

The procedure for inserting or deleting rows is exactly the same, except that you use row numbers instead of column letters.

Changing the Column Width

Column width can be changed in a couple different ways. Only one way will be described here. Output from the Descriptive Statistics Analysis Tool will be used as an example. As you can see in the output displayed below, many of the labels in column A can only be partially viewed because the column width is too narrow.

	A	B
1	TV	
2		
3	Mean	2.636364
4	Standard E	0.777791
5	Median	2
6	Mode	3
7	Standard D	2.579641
8	Sample Va	6.654545

Position the mouse pointer directly on the vertical line between A and B in the letter row at the top of the columns (A | B) so that it turns into a black plus sign. Click and drag to the right until you can read all the output labels. Release the mouse key. After adjusting the column width, your output should appear similar to the output shown below. If you have made the column too wide, you can decrease the width by clicking on the vertical line and then dragging to the left.

	A	B
1	TV	
2		
3	Mean	2.636364
4	Standard Error	0.777791
5	Median	2
6	Mode	3
7	Standard Deviation	2.579641
8	Sample Variance	6.654545

Formatting Numbers

Formatting options for numerical data are available in the Format menu. Options for displaying decimals and commas are available as well as many special format categories, such as currency, accounting, date, time, and percentage. In this section, I will explain decimals, currency, and percentage.

The same worksheet will be used for all the examples. Start with a new worksheet. Enter the numbers shown below in cells A1 through A6, and then copy them into columns B, C, and D.

*You can clear everything from a sheet by marking the whole sheet (Press **CTRL+A**) and then pressing the **Delete** key. Or you can simply move to another sheet by clicking on a sheet tab at the bottom of the screen.*

	A	B	C	D
1	1.234567	1.234567	1.234567	1.234567
2	2.34567	2.34567	2.34567	2.34567
3	3.4567	3.4567	3.4567	3.4567
4	4.567	4.567	4.567	4.567
5	5.67	5.67	5.67	5.67
6	6.7	6.7	6.7	6.7

Decimal Points

1. Click on the **B** at the top of the second column to mark it.

	A	B	C	D
1	1.234567	1.234567	1.234567	1.234567
2	2.34567	2.34567	2.34567	2.34567
3	3.4567	3.4567	3.4567	3.4567
4	4.567	4.567	4.567	4.567
5	5.67	5.67	5.67	5.67
6	6.7	6.7	6.7	6.7

2. Click **Format** in the Menu Bar and select **Cells**.

3. Click the **Number** tab at the top of the Format Cells dialog box so that the number options are displayed. Select the **Number** Category and change decimal places to 4. Click **OK**.

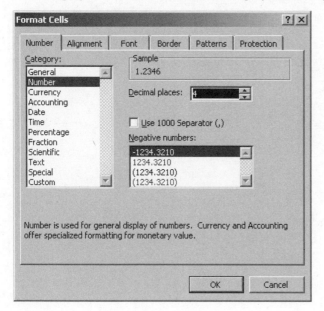

If a string of #s is displayed in a cell, that indicates that the number is too long to be displayed with the current column width. You can adjust the column width by double-clicking between the B and C at the top of the column. Excel will automatically increase the column width to accommodate the longest entry.

Currency

1. Click on **C** at the top of the third column to mark it**.**

	A	B	C	D
1	1.234567	1.2346	1.234567	1.234567
2	2.34567	2.3457	2.34567	2.34567
3	3.4567	3.4567	3.4567	3.4567
4	4.567	4.5670	4.567	4.567
5	5.67	5.6700	5.67	5.67
6	6.7	6.7000	6.7	6.7

2. Click **Format** and select **Cells**.

3. Click the **Number** tab at the top of the Format Cells dialog box so that the number options are displayed. Select the **Currency** Category and select **2** Decimal places. Click **OK**.

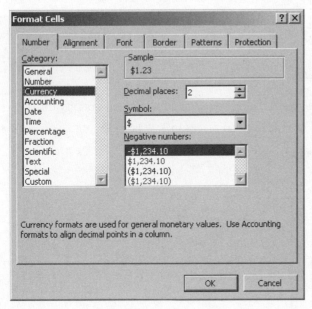

Percentage

1. Click on **D** at the top of the fourth column to mark it.

	A	B	C	D
1	1.234567	1.2346	$1.23	1.234567
2	2.34567	2.3457	$2.35	2.34567
3	3.4567	3.4567	$3.46	3.4567
4	4.567	4.5670	$4.57	4.567
5	5.67	5.6700	$5.67	5.67
6	6.7	6.7000	$6.70	6.7

2. Click **Format** and select **Cells**.

3. Click the **Number** tab at the top of the Format Cells dialog box so that the number options are displayed. Select the **Percentage** Category. Click **OK**.

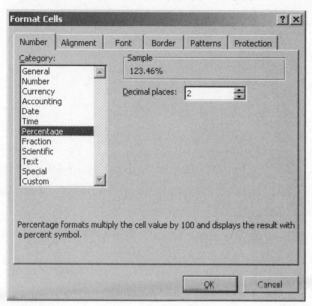

Note that the percentage format multiplies the cell value by 100. This format would be especially useful if your original cell entries were proportions, such as .123, .234, etc., and you wished to express them as percentages.

	A	B	C	D
1	1.234567	1.2346	$1.23	123.46%
2	2.34567	2.3457	$2.35	234.57%
3	3.4567	3.4567	$3.46	345.67%
4	4.567	4.5670	$4.57	456.70%
5	5.67	5.6700	$5.67	567.00%
6	6.7	6.7000	$6.70	670.00%

Recoding

Recoding refers to creating a new categorical variable for an existing variable that takes on numerical values. For example, you may have recorded age in years and now want to analyze age as a categorical variable with categories of young, middle, and older.

1.   Enter the Age values shown below. These values represent number of years.

	A
1	Age
2	18
3	38
4	21
5	45
6	52
7	60
8	68
9	37
10	40
11	35
12	29
13	30

2.   You want to recode Age into categorical variable where less than 30 years will be *Young*, 30 to 59 years will be *Middle*, and more than 59 years will be *Older*. Create a new variable with a recognizable label such as Age Group. Enter the label **Age Group** in cell **B1**.

	A	B
1	Age	Age Group
2	18	

3.   You will use the LOOKUP function to give values to Age Group. Click in cell **B2**.

4.   Enter this LOOKUP function in cell B2: =LOOKUP(A2,{0,30,60},{"Young","Middle","Older"}). Press [**Enter**].

This LOOKUP function says to take the value in cell A2 and compare it to the numbers in brackets. If the number is equal to or greater than 0 and less than 30, then assign the "Young" category. If the number is equal to or greater than 30 and less than 60, then assign the "Middle" category. If the number is equal to or greater than 60, then assign the "Older" category.

	A	B	C	D	E	F
1	Age	Age Group				
2	18	=LOOKUP(A2,{0,30,60},{"Young","Middle","Older"})				
3	38					

5. Copy the LOOKUP function in cell B2 to cells B3 through B13.

	A	B
1	Age	Age Group
2	18	Young
3	38	Middle
4	21	Young
5	45	Middle
6	52	Middle
7	60	Older
8	68	Older
9	37	Middle
10	40	Middle
11	35	Middle
12	29	Young
13	30	Middle

Sorting

Sorting is a useful tool for becoming familiar with a data set that will be analyzed. You can sort data in ascending or descending order. You also can sort by two or more variables.

1. Enter the data as displayed in the worksheet below.

	A	B	C
1	Name	Weight	Eyes
2	Fred	200	Blue
3	Sue	120	Brown
4	John	185	Brown
5	Jane	98	Blue

2. Mark the range **A1 to C5**.

	A	B	C
1	Name	Weight	Eyes
2	Fred	200	Blue
3	Sue	120	Brown
4	John	185	Brown
5	Jane	98	Blue

3. Click **Data** in the Menu Bar and select **Sort**.

4. Complete the Sort dialog box as shown below. An explanation of the entries is given immediately following the dialog box.

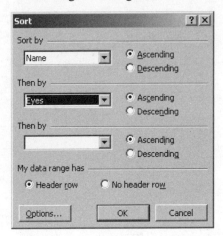

- **Sort by**. The upper left cell (Name) is listed as the default Sort by variable. If you click on the down arrow to the right of Name, you could select one of the other variables in the data set as your sorting variable. Let's use **Name** for the first sorting variable. You have the choice of sorting in ascending or descending order. Let's sort in **Ascending** order.

- You can sort by more than one variable. In the first Then by window, select **Eyes** to be sorted in **Ascending** order.

- At the bottom of the dialog box, you have to indicate whether the row at the top of the range contains data to be included in the sort, or if these are variable labels and should be treated as headers. For this example, these should be treated as headers, so **Header row** should be selected.

5. Click **OK**, and the data will be sorted as shown in the worksheet below.

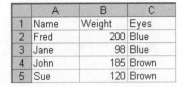

Saving and Retrieving Information

Naming Workbooks

Earlier versions of Excel used one sheet as the basic unit that was stored as one file. Now, up to 256 sheets can be stored as one unit, called a **workbook**. If you have opened a new workbook and entered data that you want to save, the next step is to click on **File** in the Menu Bar and select **Save**. If the sheet is new and not yet given a file name, you will be prompted to provide a name. The default name of **Book1** appears in the **File Name** window of the **Save As** dialog box. Unless you indicate otherwise, the file will be given the name

Book1 and will be saved in the location last used, be it the hard drive, network server, or diskette. A common practice is to save in a folder called **My Documents**. The Save As dialog box would appear as shown below.

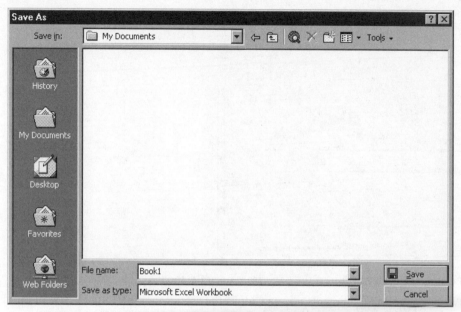

File names can have up to 218 characters. There are restrictions on the symbols that you can use in a file name, which will be apparent when you get an error message for using a / or some other forbidden symbol. Unless you type a different extension, Excel will automatically attach **xls** to an Excel file name. If you prefer a different extension, it can have a maximum of three characters. I recommend using the xls extension. Some examples of possible file names are shown below.

Opinion survey

WISCONSIN DATA

2004

1999 Japan & China Info

Printing

Page Setup

The **Page Setup** dialog box has four tabs along the top to access the four setup options: **Page**, **Margins**, **Header/Footer**, and **Sheet**. The dialog box is accessed by clicking **File** in the Menu Bar and selecting **Page Setup**.

Page

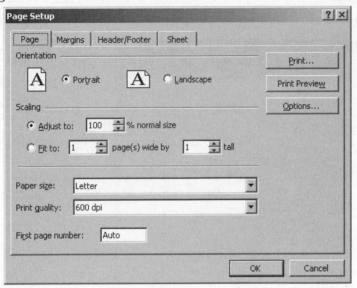

Page allows the user to change the **orientation** of the page from portrait (the way this book is printed) to landscape (sideways). With the **scaling** commands you can adjust the size of the image to be 10% of normal size or enlarge it up to 400%. The **Fit to** command instructs the program to automatically adjust the size to fit any of a number of pages as specified.

Paper Size and **Print Quality** provide options shown via the drop down arrow, although letter-size paper is used most often. On some printers, you can select a higher-quality print by changing the dpi to a value of 600 or more. This does increase printing time and use of ink, however.

Clicking on **Options** provides more graphic views of some of these choices, depending on your printer.

First Page Number is set to **Auto** as the default. If you wish to start the page numbering at a specific value, type in that value after highlighting **Auto**.

Margins

Click on the **Margins** tab to choose margin options.

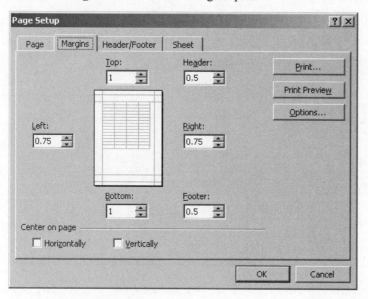

The preview window shows how the margins will look as you change them. Note that you can also center the output on the page horizontally, vertically, or both. After you have changed any of the settings, you can click on **Print Preview** to check the worksheet's appearance. You can also change the margins in Print Preview by dragging the margin boundary handles.

Headers and Footers

Headers contain the information printed across the top of all (or all but the first) pages. Information that is printed at the bottom of a page is called a **Footer**.

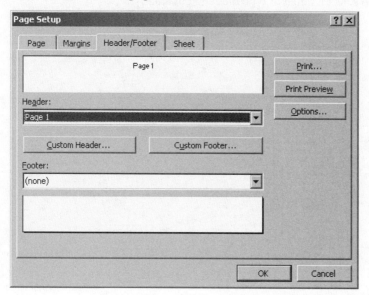

Click on the arrow to the right of the Header window, and you will see several standardized options for the **header**. I selected the Page 1 option. Similarly, several standardized options are available for the **footer**. If you wish, you can click on **Custom Header** or **Custom Footer** and type the information in the spaces provided. The Header dialog box is shown below.

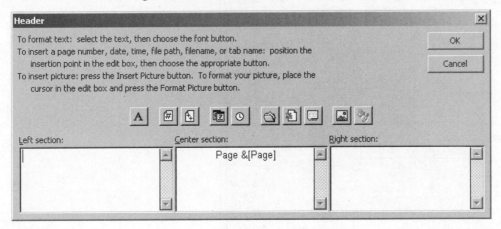

You can place your header in the left, center, or right of the page. Or you can use all three if you like. Refer to the information at the top of the dialog box for brief descriptions of the icons. For example, if you click the letter A, you will be able to specify the font, font style, and size for selected text.

Sheet

The sheet dialog box allows you to choose the area to print, column titles, row titles, gridlines, and print quality.

Print area allows you to select a particular range to print. Simply click in the Print area window and then move to the worksheet. Click and drag to select a range or type the range manually.

Print titles allows you to select rows and/or columns of the worksheet that will appear on all pages. In the Page Setup dialog box shown here, I have requested that the first row appear on all pages.

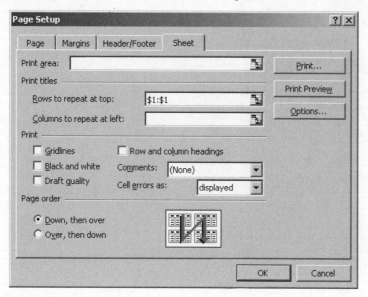

The **Print** option most often selected is **gridlines**. The output shown below on the left was printed without gridlines and the output on the right was printed with gridlines. As you can see, the gridlines make data set easier to read.

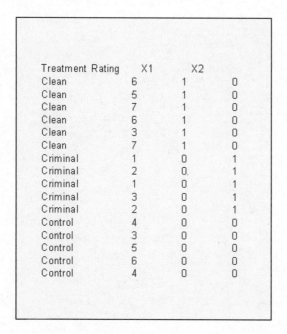

To select gridlines, click in the box to the left of **Gridlines** as shown below. Then click **OK**.

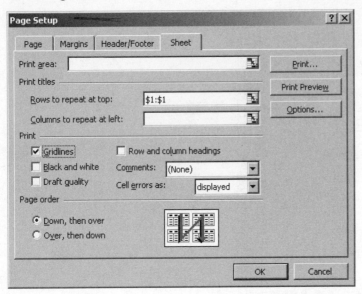

Chapter 3

Formulas

Operators

Algebraic formulas utilize the four common mathematical operations of addition (+), subtraction (-), multiplication (x), and division (÷). When creating formulas, we will use these four operations and three others: negation, exponentiation, and percent. In Excel, these **operators** are symbolized as follows:

Addition +

Subtraction –

Multiplication *

Division /

Negation, using the minus sign to indicate a negative number, as in -3

Exponentiation ^

Percent %

Multiplication is indicated by an asterisk (*) rather than an x. Division is indicated by a diagonal (/) instead of the division symbol. If we wish to raise a value to a power, say, X squared, we place a carat (^) between the X and the 2. In Excel, the formula for X squared appears as X^2. We can place a percent sign (%) after a value, as in 25%. For example, the formula **=15^2*25%** raises 15 to the second power and multiplies the result by .25 (the decimal form of 25%) to produce the result of 56.25.

Operations can be grouped together within parentheses. The parentheses determine the order in which the commands will be executed by Excel. Any operations enclosed in parentheses are executed first, from the innermost parentheses to the outermost.

Order of Operations

The order of operations in Excel is:

1. Negation
2. Percent
3. Exponentiation
4. Multiplication and division
5. Addition and subtraction

Excel first calculates expressions in parentheses and then uses those results to complete the calculations of the formula. Examples are given at the top of the next page.

=2+4*3   produces 14 because multiplication occurs before addition.

=(2+4)*3   produces 18 because operations within parentheses are executed first.

=2*(4+3)   produces 14 because operations within parentheses are executed first.

In a more complex formula,

=2+(4*(3+5)^2)/2   produces 130.

The more complex formula is carried out as follows: First, the terms in the innermost parentheses are executed, producing 8. Next, the 8 is raised to the second power, producing 64, since this is within the second set of parentheses and exponentiation has precedence over the other operations. Then 64 is multiplied by 4, yielding 256. The 256 is then divided by 2, producing 128. Finally, 2 is added to 128, yielding 130. Try this on your computer by keying the formula into cell A1. Press [**Enter**] when the formula is complete.

Writing Equations

In the equations above, we always used numeric constants such as 2, 3, 4, or 5. With Excel, the formulas that you write will more likely contain an address for a cell that can hold most any value. This is equivalent to the X and Y that act as unknowns in algebra. For example, to convert degrees Fahrenheit to degrees Celsius, subtract 32 from the Fahrenheit temperature, multiply by 5, and divide by 9. In Excel, this formula would begin with an equal sign and would appear as =**(F-32)*5/9**, where F is the Fahrenheit temperature that we wish to convert to Celsius. Let's set up a table to do this conversion.

1. Type the label **Fahrenheit** in cell **A1**. Type **Celsius** in **B1.**

2. Enter the two values of **65** and **64** in cells **A2** and **A3** as shown in the worksheet below.

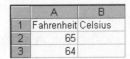

3. Let Excel continue the Fahrenheit series. Click on **65** and drag down to cell **A80** to mark those cells. Click **Edit** in the Menu Bar. Select **Fill→Series**.

4. Notice that Excel has correctly inferred that the step size is –1. We want to continue the series using that one-unit change. Click **OK** and the temperatures extend from 65 to –13 degrees Fahrenheit.

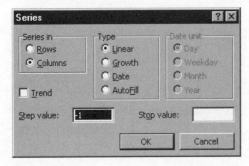

5.   Let's calculate degrees Celsius using the conversion formula and place the results in column B. Activate cell **B2** by clicking in it.

6.   **Entering a formula**. We will use A2 in the formula. A2 is the cell address of 65 degrees Fahrenheit. Enter the formula **=(A2-32)*5/9**.

	A	B	C
1	Fahrenheit	Celsius	
2	65	=(A2-32)*5/9	
3	64		

7.   Press [**Enter**]. You should now see 18.3333 in cell B2.

	A	B
1	Fahrenheit	Celsius
2	65	18.33333
3	64	

8.   **Copying a formula with a cell address**. Next we will copy the formula in cell B2 down to cell B80. Activate cell **B2**. Then click on the small square in the lower right of cell B2 (the fill handle) and drag to cell B80. The screen pointer becomes a solid black plus sign when you use the fill handle. If these steps have been carried out correctly, your worksheet should look like the one shown below. The address of each A column cell that contained degrees Fahrenheit was changed in the formula as it was copied into the cells of column B. Click in cell B3 and look in the Formula Bar to check the contents of the cell. You should see =(A3-32)*5/9.

B3	▼	fx	=(A3-32)*5/9	
	A	B	C	D
1	Fahrenheit	Celsius		
2	65	18.33333		
3	64	17.77778		
4	63	17.22222		
5	62	16.66667		
6	61	16.11111		
7	60	15.55556		
8	59	15		
9	58	14.44444		
10	57	13.88889		
11	56	13.33333		
12	55	12.77778		
13	54	12.22222		
14	53	11.66667		
15	52	11.11111		

9.   **Formatting numbers**. To make our output more attractive, let's display all the Celsius values with two decimal places. Click on letter **B** at the very top of column B to activate that column. Click **Format** in the Menu Bar and select **Cells**.

10. In the Format Cells dialog box, select the **Number** Category and look at the number displayed in the Decimal places window. You want 2 decimal places. If necessary, use the arrows to the right of the window to change the value to 2.

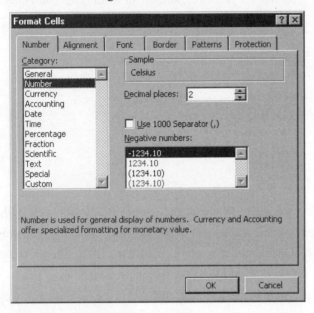

11. Click **OK**. The first several rows of your worksheet should now appear like the ones shown below.

	A	B
1	Fahrenheit	Celsius
2	65	18.33
3	64	17.78
4	63	17.22
5	62	16.67
6	61	16.11
7	60	15.56
8	59	15.00

Relative References

When we entered the formula for determining Celsius degrees in cell B2 for our temperature conversion example, we entered it using the relative reference mode. We told the program to go to cell A2 and obtain the value located there to use in the formula. Actually, we instructed Excel to move one cell location to the left of the active cell and obtain the value that was stored there. So, when the formula in B2 was copied to B3, Excel moved one cell to the left to obtain the Fahrenheit value in A3, and so on. In other words, when the cell address is relative, Excel moves to a cell address relative to its starting position. Relative addresses are written without dollar signs. For example, B2 is a relative address.

Absolute References

Regardless of the starting point, absolute references do not change. For example, let's say that we are given the monthly sales of candy bars at three convenience stores: $300, $560 and $725. We want to know the average daily sales. If there are 30 days in each of these months, we would divide monthly sales by 30 to find the daily average. In this example, 30 is a constant, meaning that it does not change. Similarly, if the cell address of 30 were used instead of the actual value, the computing formula would need to express the cell address in such a way that would indicate that it does not change. In other words, the cell address needs to be absolute. Absolute cell addresses are written with dollar signs. For example, $B$2 is an absolute address. Work through this short example to see how to use an absolute address.

1. Enter the monthly sales of **300**, **560**, and **725** in the first three cells of column A as shown below, and enter **30** in cell A6.

	A
1	300
2	560
3	725
4	
5	
6	30

2. Now let's compute average daily sales using a formula. Activate cell **B1**. Key in this formula to compute daily sales: **=A1/$A$6**.

The period at the end of the sentence is not part of the formula.

	A	B
1	300	=A1/$A$6
2	560	
3	725	
4		
5		
6	30	

3. Press [**Enter**]. You should now see 10 in cell B1.

	A	B
1	300	10
2	560	
3	725	
4		
5		
6	30	

4. Let's copy the formula in B1 to cells B2 and B3. Activate cell **B1**. Then click on the small square in the lower right of cell B1 (the fill handle) and drag to cell B3. As you can see, the values in cells A2 and A3 were each divided by 30.

	A	B
1	300	10
2	560	18.66667
3	725	24.16667
4		
5		
6	30	

5. Click in cell B2 and look at the Formula Bar. You will see A2 and $A$6. Then click in cell B3 and look at the Formula Bar. You will see A3 and $A$6. The first cell reference is relative—it changes

when the formula is copied down from one row to another. The second reference is absolute—it does not change.

Depending on the nature of the calculation you need to make, you can make both the column and row of a cell reference absolute (e.g., $A$1), only the column (e.g., $A1), or only the row (e.g., A$1).

Using Formulas in Statistics

For beginning statistics students, many analyses are carried out using a hand-held calculator. These calculations could also be performed using Excel. An important advantage associated with the use of Excel is that the student will know whether or not a number was entered correctly just by looking at the worksheet. Another important advantage is that the student will be able to carry out calculations quickly and accurately by copying a formula in one cell to other cells. In this section, I will present instructions for using formulas to calculate the mean, deviation scores, squared deviation scores, variance, standard deviation, and Z-scores. The sections are presented in the same order in which the calculations would be carried out if you were doing them by hand. If you are using Excel to do the examples shown in this section of the manual, you will need to start with mean, then deviation scores, then squared deviation scores, and so on, ending with Z-scores.

Sample Research Problem

A researcher was interested in describing the classroom behavior of gifted high school students. In particular, the researcher was interested in the number of times that the gifted students volunteered to answer questions or make comments in a science class. At the end of one class session, the five gifted students enrolled in the course received the participation scores shown below.

	A	B
1		Score
2		3
3		14
4		0
5		17
6		6

Mean

1. Enter the five participation scores in column B of an Excel worksheet as shown above. We'll use column A for labels.

2. We will begin by calculating the sum of the five scores. First, activate cell **A7** and key in **Sum**.

	A	B
1		Score
2		3
3		14
4		0
5		17
6		6
7	Sum	

3. Click in cell **B7**. Then click the AutoSum button near the top of your screen. The AutoSum button is the Greek Sigma, 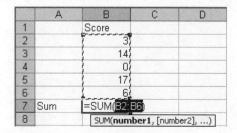, found on the Standard Toolbar. AutoSum automatically enters the equal sign, the SUM function, and the range of numerical values immediately above the activated cell.

Instead of using the AutoSum button, you could type =Sum.

	A	B	C	D
1		Score		
2		3		
3		14		
4		0		
5		17		
6		6		
7	Sum	=SUM(B2:B6)		
8		SUM(**number1**, [number2], …)		

4. Check to make sure that the range is accurate. If it is not, make the necessary corrections. When the range is correct, press [**Enter**]. You should see 40 in cell B7.

	A	B
1		Score
2		3
3		14
4		0
5		17
6		6
7	Sum	40

5. The mean is computed by dividing the sum by the number of observations. Let's use Count to refer to the number of observations. Activate cell **A8** and key in **Count**.

6. Activate cell **B8**. We will now use Excel's COUNT function to provide the number of observations.

	A	B
1		Score
2		3
3		14
4		0
5		17
6		6
7	Sum	40
8	Count	

7. At the top of the screen, click **Insert** and select **Function**.

8. In the Insert Function dialog box, select the **Statistical** category and the **COUNT** function. Click **OK**.

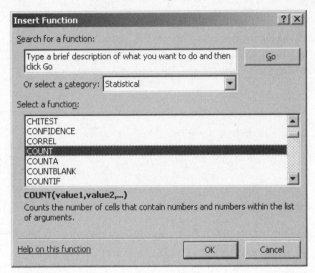

9. In the Function Arguments dialog box, you should see B2:B7. This needs to be changed to B2:B6, because you do not want to include the sum in the count.

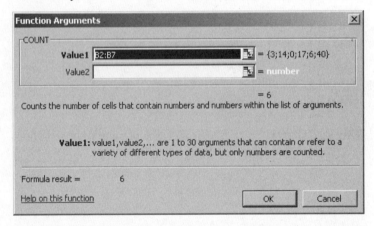

10. Click in the **Value 1** window and change B7 to **B6**. Click **OK**.

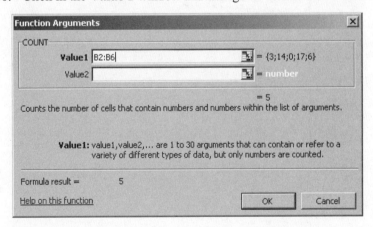

11. You should now see 5 in cell B8. Activate cell **A9** and type **Mean**.

	A	B
1		Score
2		3
3		14
4		0
5		17
6		6
7	Sum	40
8	Count	5
9	Mean	

12. Activate cell **B9**. To compute the mean, we divide Sum by Count. Using cell addresses, this formula would be expressed as =B7/B8. The recommended procedure is to click in the cell addresses to make them part of the formula. Type = in cell B9 and then click in cell **B7**. Type / and then click in cell **B8**. In the Formula Bar, you should see =B7/B8.

	A	B
1		Score
2		3
3		14
4		0
5		17
0		6
7	Sum	40
8	Count	5
9	Mean	=B7/B8

13. Press [**Enter**]. You should now see a value of 8 in cell B9.

You could also type the cell addresses to enter them in formulas, but you will find that it is more accurate and faster to click in the cells.

	A	B
1		Score
2		3
3		14
4		0
5		17
6		6
7	Sum	40
8	Count	5
9	Mean	8

Deviation Scores

1. Activate cell **C1** and enter the label **Dev Score**.

2. Activate **C2**. To calculate a deviation score, we subtract the mean from the raw score. For the first score in the data set, this formula would be expressed as =B2-B9. Type = in cell C2. Then click in cell **B2**. Type - and then click in cell **B9**. Press [**Enter**] and you see –5 in cell C2.

	A	B	C
1		Score	Dev Score
2		3	=B2-B9
3		14	
4		0	
5		17	
6		6	
7	Sum	40	
8	Count	5	
9	Mean	8	

3. Dollar signs need to be added to B9 to make it an absolute reference. Click in **C2**. In the Formula Bar near the top of the screen, you should see =B2-B9.   Move your cursor to the Formula Bar and click and drag over **B9** so that it is highlighted.

COUNT	▾ ✗ ✓	*fx* =B2-B9		
	A	B	C	D
1		Score	Dev Score	
2		3	=B2-B9	
3		14		
4		0		
5		17		
6		6		
7	Sum	40		
8	Count	5		
9	Mean	8		

4. Press the **F4** key (usually located near the top left side of the keyboard). The cell reference should change to $B$9. Press [**Enter**]. Now, when you copy the formula in cell C2, each score will be divided by the value in cell B9.

The dollar signs could also be entered manually.

COUNT	▾ ✗ ✓	*fx* =B2-$B$9		
	A	B	C	D
1		Score	Dev Score	
2		3	=B2-$B$9	
3		14		
4		0		
5		17		
6		6		
7	Sum	40		
8	Count	5		
9	Mean	8		

5. Copy the formula in C2 to cells C3 through C6. Your worksheet should now look like the one shown below.

	A	B	C
1		Score	Dev Score
2		3	-5
3		14	6
4		0	-8
5		17	9
6		6	-2
7	Sum	40	
8	Count	5	
9	Mean	8	

Squared Deviation Scores

1. Next, we will square each of the deviation scores. Activate cell **D1** and type **Sqd Dev Score**. Adjust the column width so that the entire label is displayed.

2. The formula that you will use to square the first deviation score is =C2^2. C2 is the cell address of the deviation score. The ^ is the exponentiation sign. Activate **D2**. Type =. Click in cell **C2**. Type ^2. Press [**Enter**]. You will see 25 in cell D2.

	A	B	C	D
1		Score	Dev Score	Sqd Dev Score
2		3	-5	=C2^2
3		14	6	
4		0	-8	
5		17	9	
6		6	-2	
7	Sum	40		
8	Count	5		
9	Mean	8		

3. Copy cell D2 to cells D3 through D6. Your worksheet should now look like the one displayed below.

	A	B	C	D
1		Score	Dev Score	Sqd Dev Score
2		3	-5	25
3		14	6	36
4		0	-8	64
5		17	9	81
6		6	-2	4
7	Sum	40		
8	Count	5		
9	Mean	8		

Variance

1. Activate cell **A10**. Type **Variance**. Then activate cell **B10** where the value of the variance will be placed.

	A	B	C	D
1		Score	Dev Score	Sqd Dev Score
2		3	-5	25
3		14	6	36
4		0	-8	64
5		17	9	81
6		6	-2	4
7	Sum	40		
8	Count	5		
9	Mean	8		
10	Variance			

2. To calculate the variance, we will use the formula for a population variance. In this formula, the sum of the squared deviation scores is divided by the number of observations. For our example, that formula is expressed as =SUM(D2:D6)/B8. To begin the formula, click on the AutoSum button $\boxed{\Sigma}$. Because Excel automatically enters the range of numerical values immediately above the active cell, you will probably see =SUM(B8:B9). You need to change this to D2:D6.

	A	B	C	D
1		Score	Dev Score	Sqd Dev Score
2		3	-5	25
3		14	6	36
4		0	-8	64
5		17	9	81
6		6	-2	4
7	Sum	40		
8	Count	5		
9	Mean	8		
10	Variance	=SUM(B8:B9)		
11		SUM(**number1**, [number2], ...)		

3. Move the cursor to the Formula Bar. Click and drag over **B8:B9** so that it is highlighted.

COUNT	▼ ✕ ✓ *fx*	=SUM(B8:B9)	
	A	B	SUM(**number1**, [number2], ...)
1		Score	Dev Score	Sqd Dev Score
2		3	-5	25
3		14	6	36
4		0	-8	64
5		17	9	81
6		6	-2	4
7	Sum	40	
8	Count	5	
9	Mean	8	
10	Variance	=SUM(B8:B9)	

4. Next, in the worksheet, click and drag over **D2** through **D6**. You will see D2:D6 in both the Formula Bar and in cell B10.

COUNT	▼ ✗ ✓ *fx*	=SUM(D2:D6|	
	A	B	SUM(**number1**, [number2], ...)
1		Score	Dev Score	Sqd Dev Score
2		3	-5	25
3		14	6	36
4		0	-8	64
5		17	9	81
6		6	-2	4
7	Sum	40		
8	Count	5		
9	Mean	8		
10	Variance	=SUM(D2:D6)		

5. To complete the formula, in the Formula Bar, position the flashing I-beam at the end of the entries—after the second parenthesis.

6. Type / to indicate division. Then click in cell **B8**, the cell address of Count.

If you want to use the unbiased formula to estimate a population variance from sample data, the denominator is N-1   Edit the formula so that it reads   SUM(D2.D6)/(B8-1).

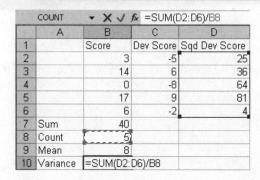

COUNT	▼ ✗ ✓ *fx*	=SUM(D2:D6)/B8		
	A	B	C	D
1		Score	Dev Score	Sqd Dev Score
2		3	-5	25
3		14	6	36
4		0	-8	64
5		17	9	81
6		6	-2	4
7	Sum	40		
8	Count	5		
9	Mean	8		
10	Variance	=SUM(D2:D6)/B8		

7. Press [**Enter**]. A variance equal to 42 is now displayed in cell B10 of the worksheet.

	A	B	C	D
1		Score	Dev Score	Sqd Dev Score
2		3	-5	25
3		14	6	36
4		0	-8	64
5		17	9	81
6		6	-2	4
7	Sum	40		
8	Count	5		
9	Mean	8		
10	Variance	42		

Standard Deviation

1. Activate cell **A11**. Type **St Dev**. Then activate cell **B11** where the value of the standard deviation will be placed.

2. The standard deviation is equal to the square root of the variance. We will use Excel's square root function, SQRT. Type =. Then click **Insert** in the Menu Bar and select **Function**.

Instead of using Insert, you could type =SQRT. Both upper and lower case are accepted.

3. In the Insert Function dialog box, select the **Math & Trig** category and the **SQRT** function. Click **OK**.

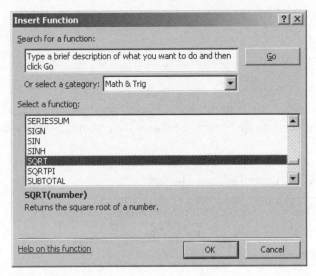

4. We want to take the square root of the variance which is located in cell B10. The flashing I-beam should be positioned in the Number window, indicating that it is ready for an entry. Click in cell **B10** of the worksheet.

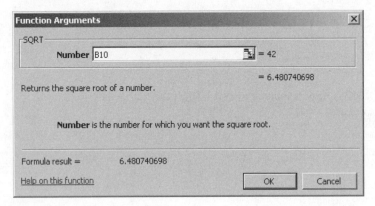

5. Click **OK**. A value of 6.480741 is now displayed in cell B11. If you increase the column width, you will see more decimal places. Note that 6.480740698 is displayed in the lower left of the dialog box shown above. By using Excel, you will not have rounding errors like those that typically occur when you carry out calculations by hand.

	A	B	C	D
1		Score	Dev Score	Sqd Dev Score
2		3	-5	25
3		14	6	36
4		0	-8	64
5		17	9	81
6		6	-2	4
7	Sum	40		
8	Count	5		
9	Mean	8		
10	Variance	42		
11	St Dev	6.480741		

Z-Scores

1. As a last step, we will express each student's score as a Z-score. Activate cell **E1** and type **Z-score**.

2. Activate **E2** where the first Z-score will be placed. The Z-score is equal to the student's deviation score (in cell C2) divided by the standard deviation (in cell B11). In the Excel worksheet, the formula is expressed as =C2/$B$11. Type =. Click in cell **C2**. Type /. Click in cell **B11**. Press [**Enter**].

	A	B	C	D	E
1		Score	Dev Score	Sqd Dev Score	Z-score
2		3	-5	25	=C2/B11
3		14	6	36	
4		0	-8	64	
5		17	9	81	
6		6	-2	4	
7	Sum	40			
8	Count	5			
9	Mean	8			
10	Variance	42			
11	St Dev	6.480741			

3. Dollar signs need to be added to B11 to make it an absolute reference. Click in **E2**. In the Formula Bar near the top of the screen, you should see =C2/B11. Move your cursor to the Formula Bar and click and drag over **B11** so that it is highlighted.

SQRT	▼ X ✓ ƒx	=C2/B11			
	A	B	C	D	E
1		Score	Dev Score	Sqd Dev Score	Z-score
2		3	-5	25	=C2/B11
3		14	6	36	
4		0	-8	64	
5		17	9	81	
6		6	-2	4	
7	Sum	40			
8	Count	5			
9	Mean	8			
10	Variance	42			
11	St Dev	6.480741			

4. Press the **F4** key. The cell reference should change to $B$11. Press [**Enter**].   Now, when you copy the formula in cell E2, each deviation score will be divided by the value in cell B11.

	SQRT	▾ ✕ ✓	fx =C2/$B$11		
	A	B	C	D	E
1		Score	Dev Score	Sqd Dev Score	Z-score
2		3	-5	25	2/$B$11
3		14	6	36	
4		0	-8	64	
5		17	9	81	
6		6	-2	4	
7	Sum	40			
8	Count	5			
9	Mean	8			
10	Variance	42			
11	St Dev	6.480741			

5. Press [**Enter**]. Copy the formula in cell E2 to cells E3 through E6. You worksheet should look similar to the one displayed below.

	A	B	C	D	E
1		Score	Dev Score	Sqd Dev Score	Z-score
2		3	-5	25	-0.77152
3		14	6	36	0.92582
4		0	-8	64	-1.23443
5		17	9	81	1.38873
6		6	-2	4	-0.30861
7	Sum	40			
8	Count	5			
9	Mean	8			
10	Variance	42			
11	St Dev	6.480741			

6. Let's display the Z-scores in column E with four decimal places. Click and drag over cells E2 through E6 so that they are highlighted.

	A	B	C	D	E
1		Score	Dev Score	Sqd Dev Score	Z-score
2		3	-5	25	-0.77152
3		14	6	36	0.92582
4		0	-8	64	-1.23443
5		17	9	81	1.38873
6		6	-2	4	-0.30861
7	Sum	40			
8	Count	5			
9	Mean	8			
10	Variance	42			
11	St Dev	6.480741			

7. Click **Format** in the Menu Bar and select **Cells**.

8. In the Format Cells dialog box, select the **Number** Category and use the up arrow to change the value in the Decimal places window to **4**.

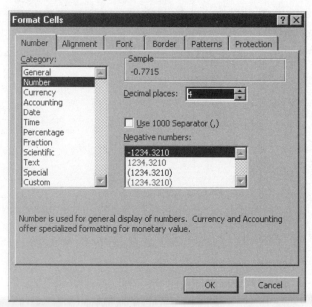

9. Click **OK**. The completed worksheet is shown below.

	A	B	C	D	E
1		Score	Dev Score	Sqd Dev Score	Z-score
2		3	-5	25	-0.7715
3		14	6	36	0.9258
4		0	-8	64	-1.2344
5		17	9	81	1.3887
6		6	-2	4	-0.3086
7	Sum	40			
8	Count	5			
9	Mean	8			
10	Variance	42			
11	St Dev	6.480741			

Because cell addresses were used in the formulas, corrections are easily made. For example, if you change the value in cell B2 from 3 to 2, you will see that all the other values will change accordingly.

10. You created a template in steps 1 through 9 that can now be used for any set of numbers. Let's use the scores **5**, **16**, **2**, **19**, and **8**. In other words, we are adding a constant of 2 points to the scores that now appear in the worksheet. Enter these new values in column B as shown below.

	A	B	C	D	E
1		Score	Dev Score	Sqd Dev Score	Z-score
2		5	-5	25	-0.7715
3		16	6	36	0.9258
4		2	-8	64	-1.2344
5		19	9	81	1.3887
6		8	-2	4	-0.3086
7	Sum	50			
8	Count	5			
9	Mean	10			
10	Variance	42			
11	St Dev	6.480741			

11. You will note that, when you add a constant, the sum and mean change, but all the other values stay the same. Let's try a different set of scores: 5, 4, 3, 2, 1.

	A	B	C	D	E
1		Score	Dev Score	Sqd Dev Score	Z-score
2		5	2	4	1.4142
3		4	1	1	0.7071
4		3	0	0	0.0000
5		2	-1	1	-0.7071
6		1	-2	4	-1.4142
7	Sum	15			
8	Count	5			
9	Mean	3			
10	Variance	2			
11	St Dev	1.414214			

12. Now, all the values, except for count, have changed. You can also use this template for a different number of scores. All you have to do is insert or delete rows. For example, click in cell **A7**. Click **Insert** in the Menu Bar and select **Rows**.

	A	B	C	D	E
1		Score	Dev Score	Sqd Dev Score	Z-score
2		5	2	4	1.4142
3		4	1	1	0.7071
4		3	0	0	0.0000
5		2	-1	1	-0.7071
6		1	-2	4	-1.4142
7					
8	Sum	15			
9	Count	5			
10	Mean	3			
11	Variance	2			
12	St Dev	1.414214			

13. A blank row 7 has been inserted. Enter these six scores in column B: **6, 5, 4, 3, 2, 1**. Because Excel copied the formulas into the inserted row, the calculations for all six scores were carried out. The completed worksheet is displayed below.

	A	B	C	D	E
1		Score	Dev Score	Sqd Dev Score	Z-score
2		6	2.5	6.25	1.8257
3		5	1.5	2.25	1.0954
4		4	0.5	0.25	0.3651
5		3	-0.5	0.25	-0.3651
6		2	-1.5	2.25	-1.0954
7		1	-2.5	6.25	-1.8257
8	Sum	21			
9	Count	6			
10	Mean	3.5			
11	Variance	1.875			
12	St Dev	1.369306			

Chapter 4

Frequency Distributions

Excel provides two means of creating a frequency distribution. One is the Pivot Table and the other is the Histogram Analysis Tool. Although Pivot Table can be used to summarize both quantitative (e.g., age) and qualitative (e.g., sex) variables, I recommend that you use the Histogram Analysis Tool for quantitative data when you want to create graph.

Frequency Distributions Using Pivot Table and Pivot Chart

Sample Research Problem

A university instructor gathered information on students enrolled in an introductory world religions course. The instructor wanted to use this information to assign students to groups for class discussion and for semester projects. We will work with the data set, shown below, that contains information on 16 students.

	A	B	C
1	SEX	AGE	RELIGION
2	M	19	Catholic
3	M	20	Jewish
4	F	22	None
5	M	23	Muslim
6	F	31	Catholic
7	F	22	Protestant
8	F	19	None
9	M	20	Jewish
10	M	19	Buddhist
11	F	20	Protestant
12	F	20	Protestant
13	M	19	Catholic
14	F	21	Protestant
15	M	19	Jewish
16	F	19	Catholic
17	F	20	Protestant

Frequency Distribution of a Quantitative Variable

I will use AGE for my example of how to create a frequency distribution for a quantitative variable.

1. Enter the student data shown above in an Excel worksheet.

2. Click on **Data** in the Menu Bar and select **Pivot Table and Pivot Chart Report**.

3. **Pivot Table and Pivot Chart Wizard-Step 1 of 3**. The data are located in an Excel worksheet, so select **Microsoft Excel List or Database**. You want to create a Pivot Table report, so select **Pivot Table**. Click **Next>**.

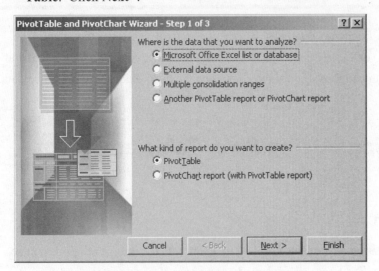

4. **Pivot Table and Pivot Chart Wizard-Step 2 of 3**. Enter the worksheet range of the AGE variable, B1 to B17. To do this, click in the **Range** window. Then click in cell **B1** of the worksheet and drag to cell **B17**. If you prefer, you can manually enter **B1:B17** in the Range window. Click **Next>**.

You could also enter the range of the entire data set, A1:C17.

5. **Pivot Table and Pivot Chart Wizard-Step 3 of 3**. Let's put the frequency distribution table in the existing worksheet, a couple of rows below the data. Click in the button to the left of **Existing worksheet** so that a black dot appears there. Then click in cell **A19** of the worksheet. If you prefer, you can manually enter **A19** in the Existing worksheet window. A19 will be the uppermost left cell of the output table. Click **Layout**.

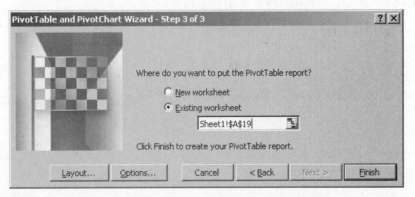

6. The diagram for creating a Pivot Table is displayed in the Layout dialog box shown below. Field buttons are displayed along the right side. For this example, we have only the AGE field button, because only the AGE column was included in the range. You will drag the field button to the desired locations in the Pivot Table. If you make a mistake, you can just drag the button off and rearrange the Pivot Table as necessary. Drag the AGE field button to **ROW**. Then drag the AGE field button to **DATA**.

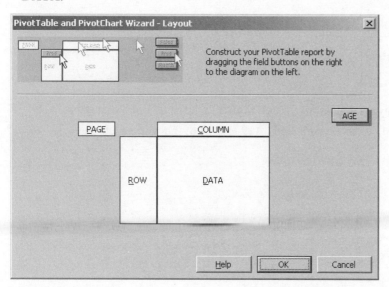

7. You will see Sum of AGE displayed in the DATA area of the Pivot Table layout. Sum is the default summary measure for a quantitative variable. For a frequency distribution, you want the summary measure called Count. To change Sum to Count, double-click on **Sum of AGE**.

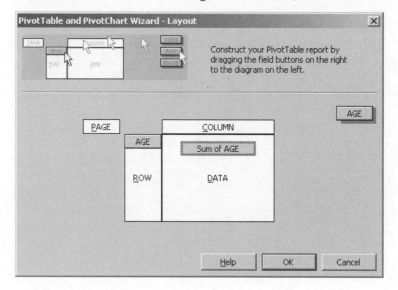

8. A Pivot Table Field dialog box like the one displayed below will appear. Click on **Count** in the list shown below **Summarize by**, and click **OK**.

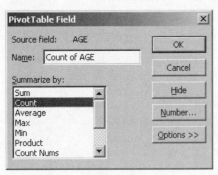

9. Click **OK** in the Pivot Table Layout dialog box. Click **Finish** in the Step 3 of 3 dialog box. The frequency distribution table will be displayed in cells A19 to B27 of the worksheet.

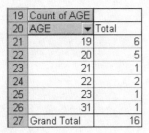

19	Count of AGE	
20	AGE ▼	Total
21	19	6
22	20	5
23	21	1
24	22	2
25	23	1
26	31	1
27	Grand Total	16

10. The AGE values in this distribution are arranged ascending order. To change to descending order, first click **AGE** in cell A20. Then click **Data** in the Menu Bar and select **Sort**. In the Sort dialog box, select **Descending**.

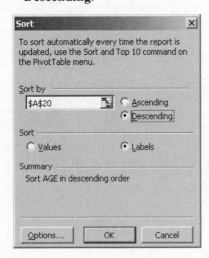

11. Click **OK** and the AGE values along with their associated frequencies will be arranged in descending order.

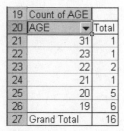

19	Count of AGE	
20	AGE ▼	Total
21	31	1
22	23	1
23	22	2
24	21	1
25	20	5
26	19	6
27	Grand Total	16

In this Pivot Table report, **Total** refers to the frequency of each age value. For example, six students were 19 years of age, and five students were 20 years of age. **Grand Total** refers to the total number of observations in the frequency distribution. Our grand total is 16—the 16 students enrolled in the world religions course.

Histogram of a Quantitative Variable

I will continue to use the AGE variable from the data presented at the beginning of this chapter to illustrate how to create a histogram of a quantitative variable. To create the histogram, you will use Pivot Chart.

1. To create a frequency graph, you need a frequency distribution. So, if you have not already done so, follow the steps in the preceding section to create a frequency distribution of the AGE variable.

2. Numbers on the horizontal axis of a frequency graph are usually arranged in ascending order. So, if the frequencies in your table are arranged in descending order, sort the values so that they are displayed in ascending order. To do this, first click **AGE** in cell A20. Then click **Data** in the Menu Bar and select **Sort**. In the Sort dialog box, select **Ascending**.

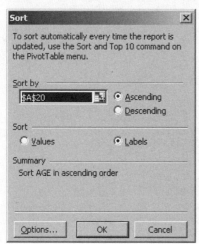

3. Click **OK**. Now that the values are displayed in ascending order in the Pivot Table report, you are ready to construct a frequency graph.

19	Count of AGE	
20	AGE ▼	Total
21	19	6
22	20	5
23	21	1
24	22	2
25	23	1
26	31	1
27	Grand Total	16

4. Select the entire Pivot Table. To do this, move the cursor in the Count of AGE cell (cell A19) so that it is pointing near the left inside edge or top inside edge and turns from a white plus sign into a black arrow. When the cursor is in the shape of a black arrow, depress the left key on the mouse. The entire table is now *selected* and will be highlighted.

19	Count of AGE	
20	AGE ▼	Total
21	19	6
22	20	5
23	21	1
24	22	2
25	23	1
26	31	1
27	Grand Total	16

5. Click **Insert** in the Menu Bar and select **Chart**. A frequency graph will be displayed in a new worksheet named Chart 1.

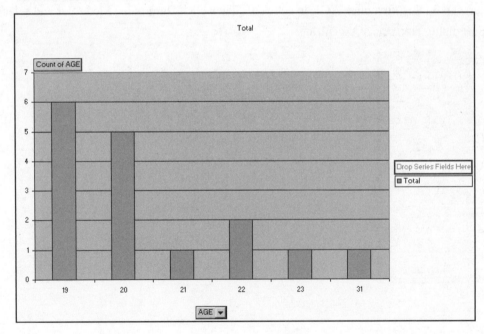

6. Now we will make a few modifications to the graph so that it will have a more attractive appearance. If the Pivot Table toolbar is not showing, click **View** and select **Toolbars → Pivot Table**.

7. First, let's get rid of the Pivot Chart field buttons. On the Pivot Table toolbar, click **Pivot Chart**. Select **Hide Pivot Chart Field Buttons**.

7. To make more modifications, we'll continue by using the Chart Wizard. Click the Chart Wizard on the Pivot Table toolbar. It looks like this: .

8. **Chart Wizard – Step 1 of 4 – Chart Type**. Select the **Column** Chart type and select the first chart in the top row under Chart sub-type. Click **Next>**.

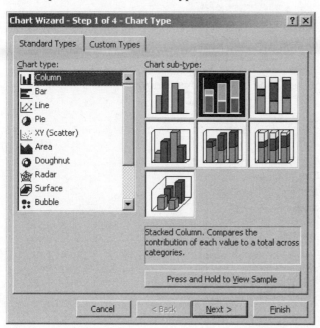

9. **Chart Wizard – Step 3 of 4 – Chart Options**. Click the **Titles** tab at the top of the Chart Options dialog box.

Step 2 of 4 asks for the worksheet location of the data. This step was omitted because you started the procedure with a completed Pivot Table report of a frequency distribution.

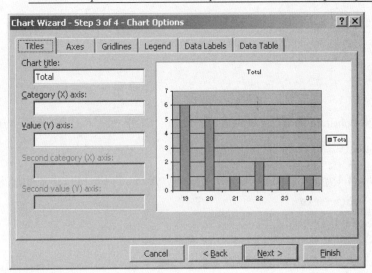

10. Enter the titles in the designated windows of the dialog box as shown below.

Chart title: **Frequencies of Age**

Category (X) variable: **Age in Years**

Category (Y) variable: **Frequency**

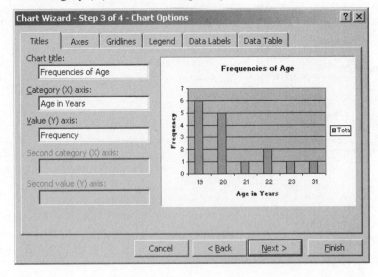

11. Click the **Legend** tab. To remove the Total legend displayed at the right of the chart, click in the **Show legend** box to remove the check mark. Click **Finish**.

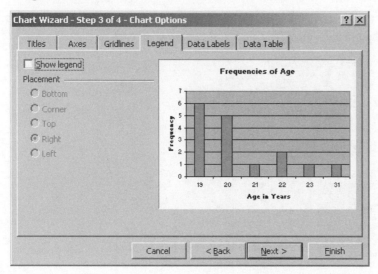

12. Let's remove the space between the vertical bars. **Right-click** on one of the vertical bars. Select **Format Data Series** from the shortcut menu that appears.

13. Click the **Options** tab at the top of the Format Data Series dialog box. Change the value in the Gap Width box to **0**. Click **OK**. Your completed graph should now look like the one displayed in the lower part of the dialog box shown below.

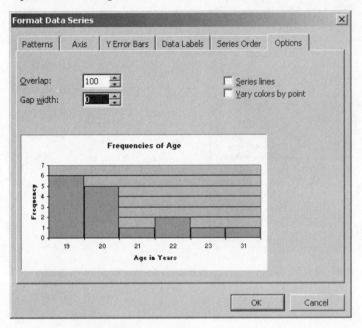

Although this is an attractive histogram, you will note that it does not accurately display the age values for this set of data because ages of 24 through 30 did not occur. You will usually want to show a frequency of zero for these values. The histogram procedure that is available in Data Analysis Tools will show these zero frequencies. Directions for the histogram tool are given later in this chapter.

Frequency Distribution of a Qualitative Variable

I will use RELIGION to illustrate how to create a frequency distribution of a qualitative variable.

1. If you have not already done so, enter the student data presented at the beginning of this chapter in an Excel worksheet. If you have already entered these data, return to the worksheet with the data by clicking on the appropriate worksheet tab at the bottom of the screen (most likely Sheet 1).

2. Click on **Data** in the Menu Bar and select **Pivot Table and Pivot Chart Report**.

3. **Pivot Table and Pivot Chart Wizard-Step 1 of 3**. The data are located in an Excel worksheet, so select **Microsoft Excel list or Database**. You want to create a Pivot Table report, so select **Pivot Table**. Click **Next>**.

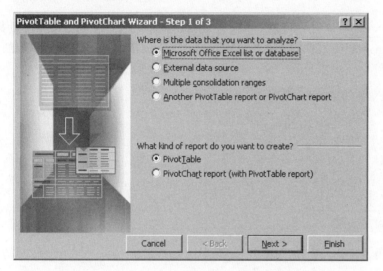

4. **Pivot Table and Pivot Chart Wizard-Step 2 of 3**. Enter the range of the RELIGION variable, C1 to C17. To do this, click in the Range window. Then click in cell **C1** of the worksheet and drag to cell **C17**. If you prefer, you can manually enter **C1:C17** in the Range window. Click **Next>**.

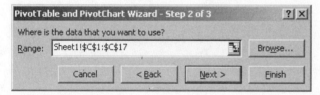

5. **Pivot Table and Pivot Chart Wizard-Step 3 of 3**. Let's put the frequency distribution in a new worksheet. Click in the button to the left of **New worksheet** so that a black dot appears there. Click **Layout** in the bottom left of the dialog box.

6. The diagram for creating a Pivot Table is displayed in the Layout dialog box. Field buttons are displayed along the right side. For this example, we have only the RELIGION button, because only the RELIGION column was included in the range. You will drag the field button to the desired locations in

the Pivot Table. If you make a mistake, you can just drag the button off and rearrange the Pivot Table as necessary. Drag the **RELIGION** field button to **ROW**. Then drag the **RELIGION** field button the **DATA**.

7. You will see Count of RELIGION displayed in the DATA area of the Pivot Table layout. Count is the default summary measure for a qualitative variable. Click **OK** in the Layout dialog box. Click **Finish** in the Step 3 of 3 dialog box.

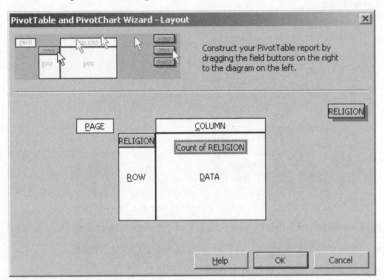

8. The frequency distribution table will be displayed in a new worksheet and should look similar to the one shown below.

9. If you want to get rid of **Drop Page Fields Here** label, just click in cell A1 where the label appears.

In this Pivot Table report, **Total** refers to the frequency of each religion. For example, one student was Buddhist and four students were Catholic. **Grand Total** refers to the total number of observations in the frequency distribution. Our grand total is 16—the 16 students enrolled in the world religions course.

Bar Graph of a Qualitative Variable

I will use RELIGION to illustrate how to create a bar graph of a qualitative variable.

1. To create a bar graph, you need a frequency distribution. So, if you have not already done so, follow the steps in the preceding section to create a frequency distribution of the RELIGION variable.

2. Select the entire Pivot Table. To do this, move the cursor in the Count of RELIGION cell (cell A3) so that it is pointing near the left inside edge or top inside edge and turns from a white plus sign into a black arrow. When the cursor is in the shape of a black arrow, depress the left key on the mouse. The entire table is now *selected* and will be highlighted.

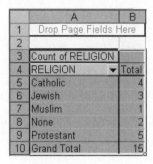

3. Click the **Chart Wizard** on the Pivot Table toolbar. It looks like this: [icon]. A frequency graph will be displayed in a new worksheet along with a Pivot Table Field List and the Pivot Table toolbar.

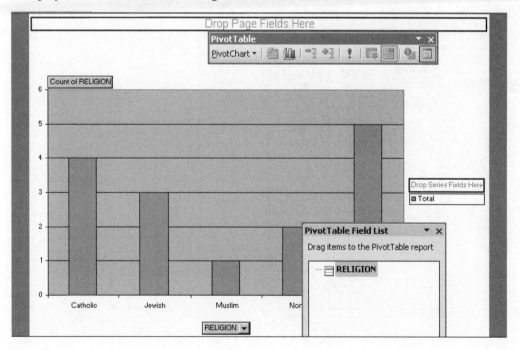

4. Now we will make a couple of modifications to the graph so it will have a more attractive appearance. You see some field buttons displayed in the graph (e.g., Count of RELIGION and Drop Series Fields Here). To remove these, click **Pivot Chart** on the Pivot Table toolbar and select **Hide Pivot Chart Field Buttons**.

5. We will use the Chart Wizard to make some more modifications. Click the **Chart Wizard** icon on the Pivot Table toolbar.

6. **Chart Wizard – Step 1 of 4 – Chart Type**. Select the **Bar** Chart type and select the first chart in the top row under Chart sub-type. Click **Next>**.

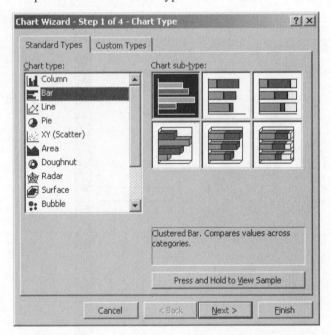

7. **Chart Wizard – Step 3 of 4 – Chart Options**. Click the **Titles** tab at the top of the Chart Options dialog box.

Step 2 of 4 asks for the worksheet location of the data. This step was omitted because you started the procedure with a completed Pivot Table report of a frequency distribution.

8. Enter a chart title and a Y-axis label in the designated windows of the dialog box as shown below.

    Chart title: **Religions of Students Enrolled in World Religions Course**

    Value (Y) axis: **Frequency**

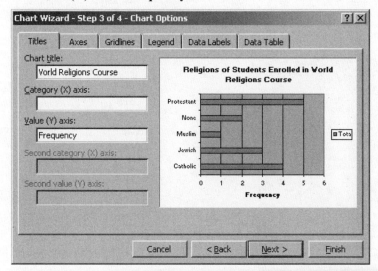

9. Click the **Legend** tab at the top of the Chart Options dialog box. To remove the Total legend displayed
    at the right of the chart, click in the **Show legend** box to remove the check mark. Click **Finish**.

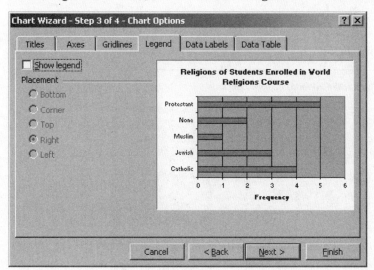

10. Let's change the numbers on the horizontal scale to display a number scale that increases by twos. Right click on a number in the horizontal scale. I clicked on 1. Then click **Format Axis**.

11. Click the **Scale** tab. Complete the entries as shown below. You are requesting that 0 be the minimum value on the number scale and that 6 be the maximum. You are requesting that the scale increase by twos (e.g., 2, 4, 6). Click **OK**.

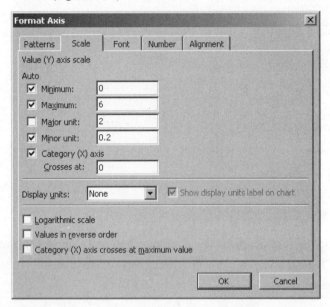

The final version of the bar graph is shown below.

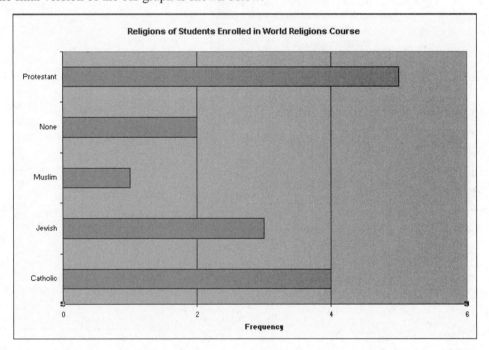

Frequency Distributions Using Data Analysis Tools

Histogram is one of the Data Analysis Tools. It provides a fairly quick means of producing both a frequency distribution and a histogram of quantitative data. The Histogram Tool can generate a table and a graph of both ungrouped and grouped data. Both of these types will be illustrated.

Sample Research Problem

A researcher developed a test that was designed to measure "cultural literacy." It consisted of multiple-choice items that asked about such things as historical events, science, art, literature, and the like. The maximum score on the test was 40.

Steps to Follow to Create a Frequency Distribution and a Histogram for Grouped Data

1. Enter these 22 cultural literacy scores in column A of an Excel worksheet as shown below.

	A	B
1	Cultural Literacy	
2	24	
3	35	
4	26	
5	29	
6	17	
7	22	
8	21	
9	17	
10	29	
11	26	
12	26	
13	16	
14	24	
15	22	
16	19	
17	25	
18	24	
19	23	
20	31	
21	14	
22	21	
23	23	

2. Excel's histogram procedure groups data and then displays them in a frequency distribution table and a frequency histogram. The procedure requires that you indicate a "bin" for each class. The number that you specify for each bin is actually the upper limit of the class. For the cultural literacy data, let's start with an upper limit of 15 and increase in increments of 3. Because 35 is the maximum score in the data set, the bins will end at 36. Enter **Bin** in cell C1 and key in the bin values as shown below.

	A	B	C
1	Cultural Literacy		Bin
2	24		15
3	35		18
4	26		21
5	29		24
6	17		27
7	22		30
8	21		33
9	17		36

3. Click **Tools** in the Menu Bar and select **Data Analysis**.

> *If Data Analysis does not appear as a choice in the Tools menu, you will need to load the Microsoft Excel ToolPak add-in. Follow the Procedure on page 7.*

4. In the Data Analysis dialog box, select the **Histogram** Analysis Tool and click **OK**.

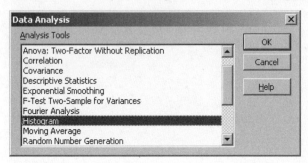

5. The Histogram dialog box will appear. Begin with the Input Range. This is the range of the Cultural Literacy scores. Click in the **Input Range** window of the Histogram dialog box. Then, in the worksheet, click and drag from cell **A1** to cell **A23**. If you prefer, you can manually enter **A1:A23** in the window.

> *The dialog box is often right on top of the data ranges that we want to mark for an analysis. You can move the dialog box by clicking on its title bar, holding down the mouse key, and dragging the box to a more convenient position.*

6. **Bin Range**. Click in the **Bin Range** window. Then click and drag from cell **C1** to cell **C9**. If you prefer, you can manually enter **C1:C9** in the window.

7. **Labels**. Click in the box to the left of **Labels** to place a check mark there. The check mark lets Excel know that the first cell in each range contains a label rather than data. If there were no check mark, Excel would attempt to use the information in these cells when constructing the frequency distribution

table and histogram. Because these cells contain words rather than data, however, you would receive an error message.

8. **Output options**. Select **New Worksheet Ply**. The output will be placed in a new worksheet with A1 as the uppermost left cell.

9. **Chart Output**. Click the **Chart Output** option so that a histogram will be produced along with the frequency distribution. The completed dialog box is shown below.

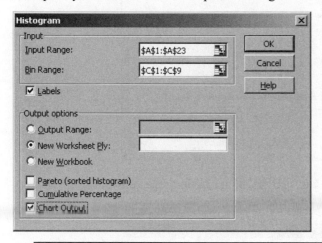

> *Click on **Help** (under **OK** and **Cancel**) to receive a detailed description of each required entry and option in the dialog box.*

10. Click **OK**. You will receive output similar to that displayed below.

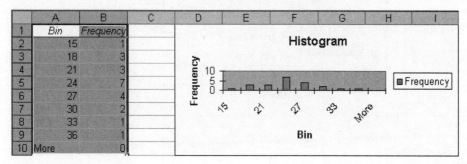

11. You will now follow steps to modify the histogram so that it is displayed in a more informative and attractive manner. First, make the chart taller so that it is easier to read. To do this, first click within the figure near a border. Black square handles appear. Click on the center handle on the bottom border of the figure and drag it down a few rows.

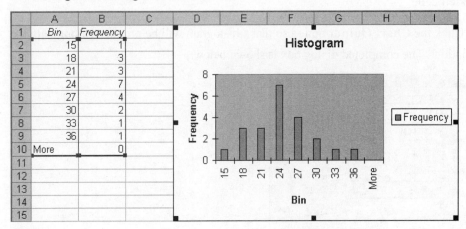

12. Next remove the space between the vertical bars. **Right-click** on one of the vertical bars. Select **Format Data Series** from the shortcut menu that appears.

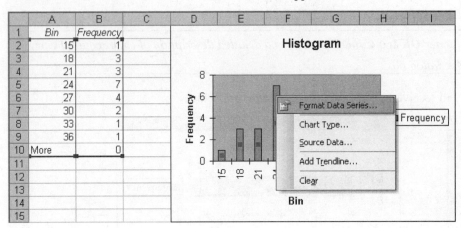

13. Click the **Options** tab at the top of the Format Data Series dialog box. Change the value in the **Gap width** window to **0**. Click **OK**.

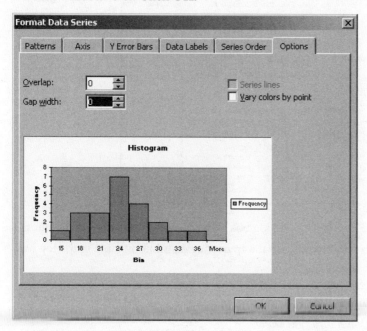

14. Change the X-axis values from upper limits to midpoints. Enter the midpoints in Column C of the Excel worksheet as shown below.

	A	B	C
1	Bin	Frequency	
2	15	1	14
3	18	3	17
4	21	3	20
5	24	7	23
6	27	4	26
7	30	2	29
8	33	1	32
9	36	1	35
10	More	0	

15. **Right-click** on a vertical bar in the chart. Select **Source Data** from the shortcut menu that appears.

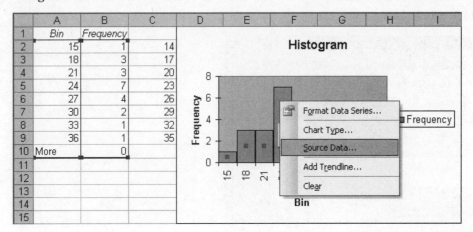

16. Click the **Series** tab at the top of the Source Data dialog box. The ranges displayed in the Values window and the Category (X) labels window refer to the frequency distribution table in the top left of the worksheet.

My frequency distribution and graph output is on Sheet 6 and that is why Sheet 6 appears in the ranges. Your output might be on a different sheet.

17. You do not want to include row 10 in the ranges, because that is the row containing information related to the "More" category and you don't want "More" printed at the end of the number scale. You also want the midpoint values in column C to be displayed on the X-axis rather than the column A bin values. First, change the 10 to 9 in the **Values** window so that the entry reads **=Sheet6!$B$2:$B$9**. Next, edit the Category (X) axis labels range so that the entry reads **=Sheet6!$C$2:$C$9**. Click **OK**.

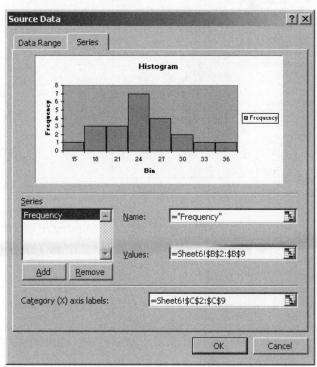

18. You will use Chart Options to modify three aspects of the histogram: Titles, gridlines, and legend. **Right-click** in the grey plot area of the chart and select **Chart Options** from the shortcut menu that appears.

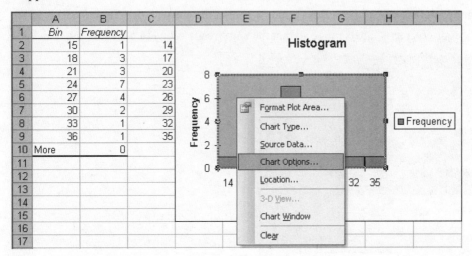

19. Click the Titles tab at the top of the Chart Options dialog box. Change the Chart title from "Histogram" to **Cultural Literacy Test Performance**. Change the Category (X) axis title from "Bin" to **Test Score**.

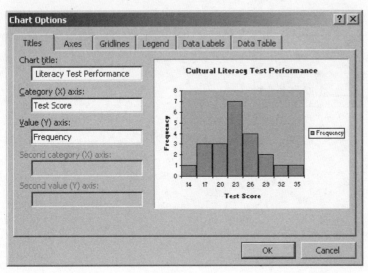

20. Click the **Gridlines** tab at the top of the Chart Options dialog box. Under Category (X) axis, click in the **Major gridlines** box so that a check mark appears there. Also, under Value (Y) axis, click in the **Major gridlines** box so that a check mark appears there.

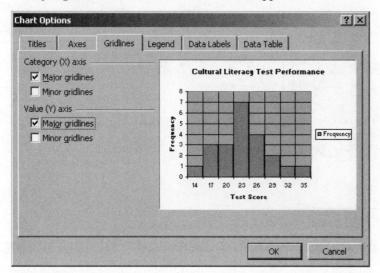

21. Click the **Legend tab**. To remove the Frequency legend displayed at the right of the histogram chart, click in the **Show legend** box to remove the check mark. Click **OK**.

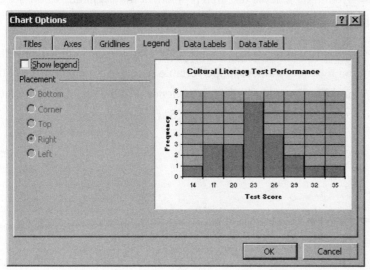

The final version of the histogram is displayed below. Several of the modifications that I presented were optional. I recommend that you experiment with your charts so that you will find the display options that best fit your needs.

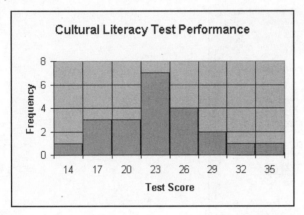

Steps to Follow to Create a Frequency Distribution and a Histogram for Ungrouped Data

1. If you have not already done so, enter the cultural literacy scores in column A of an Excel worksheet. The cultural literacy scores are displayed in the sample research problem at the beginning of the section titled Frequency Distributions Using Data Analysis Tools.

2. Excel's histogram procedure requires you to enter a "bin" for each class. When you want to display ungrouped data, you simply enter a bin for every number in the variable range. To quickly identify the highest and lowest values, just sort the data. To do this, click on the letter A at the top of the column that contains the cultural literary scores. Then click **Data** and select **Sort**.

3. The Sort dialog box should be completed as shown below. Click **OK**.

4. You can now easily see that the lowest score is 14 and the highest score is 35. Type **Bin** in cell **C1** and key in the bin values from 14 to 35 in column C as shown below.

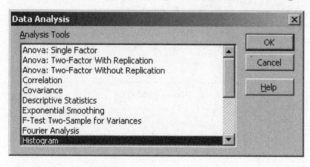

	A	B	C
1	Cultural Literacy		Bin
2	14		14
3	16		15
4	17		16
5	17		17
6	19		18
7	21		19
8	21		20
9	22		21
10	22		22
11	23		23
12	23		24
13	24		25
14	24		26
15	24		27
16	25		28
17	26		29
18	26		30
19	26		31
20	29		32
21	29		33
22	31		34
23	35		35

5. Click **Tools** in the Menu Bar and select **Data Analysis**.

6. In the Data Analysis dialog box, select the **Histogram** Analysis Tool and click **OK**.

Data Analysis

Analysis Tools

Anova: Single Factor
Anova: Two-Factor With Replication
Anova: Two-Factor Without Replication
Correlation
Covariance
Descriptive Statistics
Exponential Smoothing
F-Test Two-Sample for Variances
Fourier Analysis
Histogram

OK

Cancel

Help

7. The Histogram dialog box will appear. Enter the **Input Range** and **Bin Range** as shown below. Also
    click in the **Labels** box, the **New Worksheet Ply** button, and the **Chart Output** box.

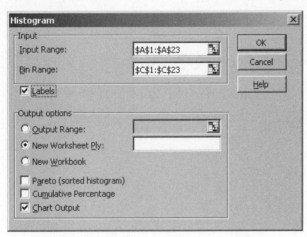

*Be sure to select **Chart Output** at the bottom of the Histogram dialog box. Otherwise you
will get a frequency distribution table and no chart.*

8. Click **OK**. You will receive output similar to the output displayed below.

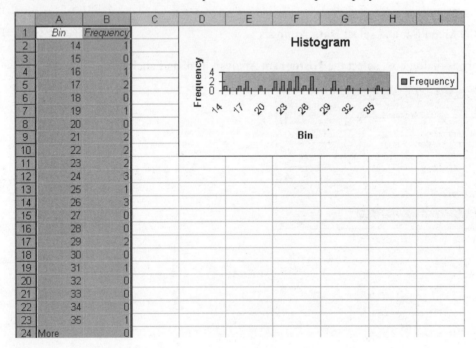

9. You will now modify the histogram. Begin by making the chart taller so that it is easier to read. To do this, first click within the figure near a border. Black square handles appear. Click on the center handle on the bottom border of the figure and drag it down a few rows.

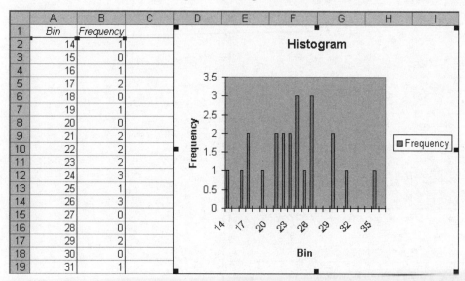

10. Next, remove the space between the vertical bars. **Right-click** on one of the vertical bars. Select **Format Data Series** from the shortcut menu that appears.

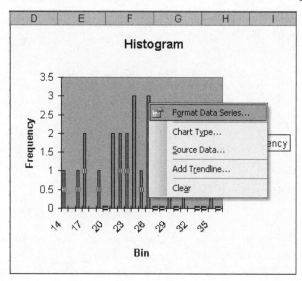

11.   Click the **Options** tab at the top of the Format Data Series dialog box. Change the value in the **Gap width** window to **0**. Click **OK**.

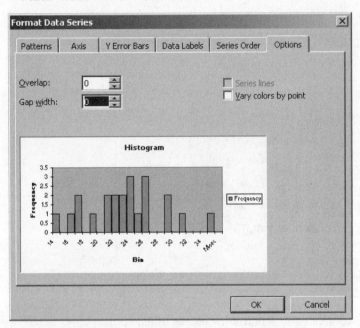

12.   Next, you will use Chart Options to modify the titles, gridlines, and legend. **Right-click** in the grey plot area of the chart and select **Chart Options** from the shortcut menu that appears.

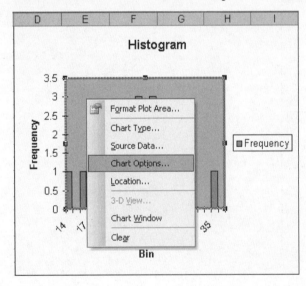

13. Click the **Titles** tab at the top of the Chart Options dialog box. Change the Chart title from "Histogram" to **Cultural Literacy Test Performance**. Change the Category (X) axis title from "Bin" to **Test Score**.

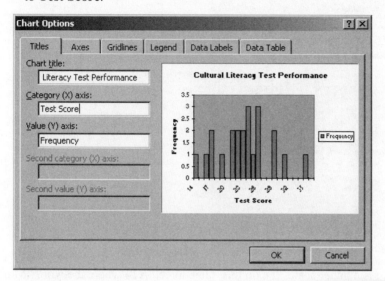

14. Click the **Gridlines** tab at the top of the dialog box. Under Category (X) axis, click in the **Major gridlines** box so that a check mark appears there. Also, under Value (Y) axis, click in the **Major gridlines** box so that a check mark appears there.

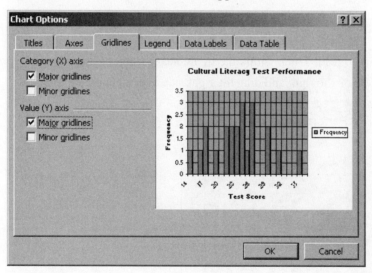

15.   Click the **Legend** tab. To remove the Frequency legend displayed at the right of the histogram chart, click in the **Show legend** box to remove the check mark. Click **OK**.

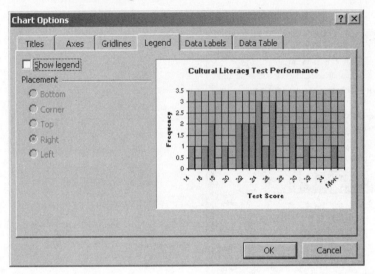

16.   Change "More" to 36 on the X-axis number scale. To do this, click in cell **A24** where the word "More" now appears and type **36**.

20	32	0
21	33	0
22	34	0
23	35	1
24	36	0

17.   Let's make the X-axis a little easier to read. Click within the figure near a border so that black square handles appear. Click on the handle on the right and drag the right side over a couple of columns.

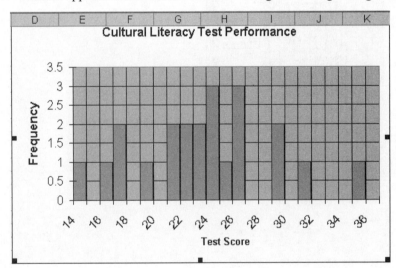

18.   **Right-click** directly on a number on the X-axis scale. I right-clicked on 24. Select **Format Axis** from the shortcut menu that appears.

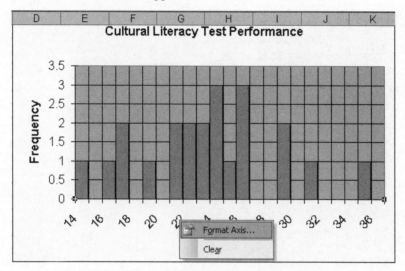

19.   Click the **Font** tab at the top of the Format Axis dialog box. Select Font Size **8**. Click **OK**.

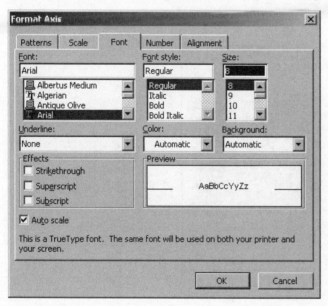

20.   Let's change the Y-axis scale so that only whole numbers appear. **Right-click** directly on one of the Y-axis numbers. I right-clicked on 1.5. Select **Format Axis** from the shortcut menu that appears.

21. Click the **Scale** tab at the top of the Format Axis dialog box. Change **Maximum** to **4**. Change **Major unit** to **1**. Click **OK**.

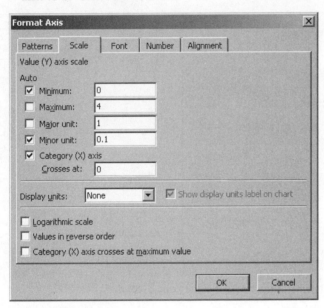

The completed histogram is shown below.

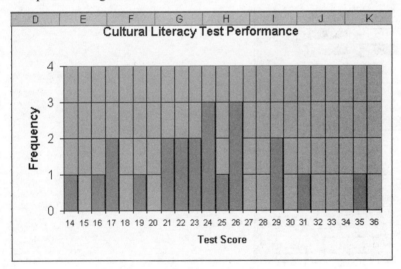

Excel provides a variety of ways to obtain summary measures for a set of data. I have included three of these in this chapter: Data Analysis Tools, Functions, and Pivot Table. I present Data Analysis Tools first because it is the simplest of the three. I end with the Pivot Table, which is easy to use but does not offer as many summary measures as Data Analysis Tools or Excel's Functions.

Data Analysis Tools: Descriptive Statistics

Data Analysis is located in the Tools menu. The Descriptive Statistics Analysis Tool provides a quick and simple way of obtaining, simultaneously, several descriptive measures for a single variable or for two or more variables. These measures include: sum, count, mean, median, mode, minimum, maximum, range, standard deviation, variance, standard error, kurtosis, and skewness.

Sample Research Problem

An electric utility company wanted to gather information regarding energy usage in its service territory. The company conducted an extensive survey that included many questions, two of which were the number of television sets and microwave ovens owned by residential customers. I have used the responses provided by 11 customers to explain how to obtain descriptive statistics utilizing the Descriptive Statistics Analysis Tool.

Steps to Follow to Obtain Descriptive Statistics for One Variable

1. Enter the data in an Excel worksheet as shown below.

	A	B
1	TV	MICROWAVE
2	3	1
3	0	1
4	9	2
5	2	1
6	3	3
7	1	0
8	2	1
9	0	2
10	3	1
11	5	1
12	1	1

2. Let's begin by obtaining descriptive statistics for TV ownership. Click on **Tools** in the Menu Bar and select **Data Analysis** in the Tools menu.

> *If Data Analysis does not appear as a choice in the Tools menu, you will need to load the Microsoft Excel ToolPak add-in. Follow the procedure on page 7.*

3. In the Data Analysis dialog box, a list of Analysis Tools is displayed. Select **Descriptive Statistics** and click **OK**.

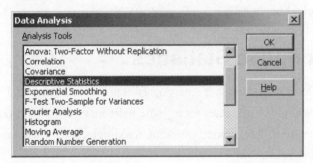

4. Complete the Descriptive Statistics dialog box as shown below. A description of each entry immediately follows the dialog box.

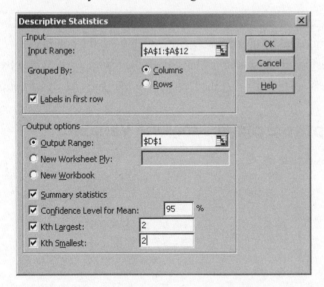

- **Input Range**. Enter the worksheet location of the data you wish to analyze. The TV data are located in A1:A12. Click in the **Input Range** window. Click in cell **A1** and then drag to cell A12. If you prefer, you can also key in the range by manually entering A1:A12. Notice that we have included A1 in the range. A1 contains "TV." TV is referred to as a label.
- **Grouped By**. Excel gives you the option of entering each variable in its own column or its own row. For this example, we used columns, which is the default option. There should be a black dot in the button to the left of **Columns**. If we had instead grouped our data by rows, we would have needed to click in the Rows button.

- **Labels in First Row**. I have found that it is very helpful to label output clearly. Therefore, I included a label in the first row for each variable. The labels will not be included in the computations but will be used to label the output. Click in the box for **Labels in First Row**.

For this example, if you forget to click Labels in First Row, you will receive the following error message: **Input range contains nonnumeric data**. *If the labels are numeric, such as ages (e.g., 8, 9, 10) or years (e.g., 1970, 1980, 1990), and you forget to click Labels in First Row, you will not receive an error message. However, the numeric labels will be included in the computations and your output will be inaccurate. An indication that this has happened is seeing labels in your output such as* **Column 1** *and* **Column 2**.

- **Output options**. Three output options are available.
  Output Range. Select Output Range if you wish to place the output in the same worksheet as the data. Enter the upper left cell of the desired location of the output table in the Output Range window. For this example, let's place the output table next to the data. Click in the **Output Range** window and then click in cell **D1** of the worksheet. If you prefer, you can manually enter **D1** in the Output Range window.
  New Worksheet Ply. This is the default option. Unless specified otherwise, the output table will be pasted into a new worksheet with A1 as the upper left cell. If you like, you can give the worksheet a special name.
  New Workbook. Select New Workbook if you want to create a new workbook to display the analysis output. The analysis output will be pasted into a worksheet of a new workbook with A1 as the upper left cell of the output table.
- **Summary statistics**. Summary statistics provided by the Descriptive Statistics Analysis Tool include the mean, standard error of the mean, median, mode, standard deviation, variance, kurtosis, skewness, range, minimum, maximum, sum, and count. Click the **Summary statistics** box so that Excel will provide these values.

If you do not select summary statistics, confidence level for mean, Kth largest, or Kth smallest, you will receive this error message: **Please choose at least one statistical option**.

- **Confidence Level for Mean**. We would like output regarding the confidence level. So, click in the box next to **Confidence Level for Mean**. The default is 95%, which is what we will use for this example. If you would like a different confidence level (e.g., 90%), enter that level in the Confidence Level for Mean window. Any number greater than 0 and less than 100 can be entered.
- **Kth Largest**. In addition to the maximum value in a data set, we might also want to know, for example, the second largest or third largest value. If so, we would click in the box next to Kth Largest and then enter the desired value in the Kth Largest window. For this example, let's ask for the second largest value. Click in the **Kth Largest** box and enter **2** in the window. The default is 1, which is the maximum value.

- **Kth Smallest**. In addition to the minimum value in a data set, we might also be interested in knowing another small value such as the second smallest. To obtain this value, click in the box next to **Kth Smallest** and then enter **2** in the window.   The default is 1, which is the minimum value.

5. Click **OK**, and the output table will be displayed with D1 as the upper left cell as requested. Adjust the column width as necessary to display all labels.

Interpreting the Output

The Descriptive Statistics output table is shown below in columns D and E. An explanation of each of the measures that is included in Descriptive Statistics is presented immediately after the output table.

	A	B	C	D	E
1	TV	MICROWAVE		TV	
2	3	1			
3	0	1		Mean	2.636364
4	9	2		Standard Error	0.777791
5	2	1		Median	2
6	3	3		Mode	3
7	1	0		Standard Deviation	2.579641
8	2	1		Sample Variance	6.654545
9	0	2		Kurtosis	3.212809
10	3	1		Skewness	1.602855
11	5	1		Range	9
12	1	1		Minimum	0
13				Maximum	9
14				Sum	29
15				Count	11
16				Largest(2)	5
17				Smallest(2)	0
18				Confidence Level(95.0%)	1.733026

- **Mean**. The arithmetic average of the number of TVs owned by the 11 residential customers in the electric utility's sample, $\overline{X} = 2.6364$ TVs.

- **Standard Error**. Standard error of the mean, $S_{\overline{X}}$, calculated using the formula $\dfrac{S}{\sqrt{n}}$, where S is the sample standard deviation and n is the number of observations. For the TV ownership distribution, $S_{\overline{X}} = 0.7778$.

- **Median**. The observation that splits the distribution in half. To determine the value of the median, the observations are first arranged in either ascending or descending order. If the number of observations is even, the median is found by taking the arithmetic average of the two middle observations. If the number of observations is odd, then the median is the middle observation. When arranged in ascending order, our sample data would be: 0, 0, 1, 1, 2, 2, 3, 3, 3, 5, 9. Because the number of observations is odd (n=11), the median is the middle (6th) observation, which is equal to 2.

- **Mode**. Observation value associated with the highest frequency. For this example, the mode is equal to 3.

A word of caution is in order regarding the value reported for the mode. Three situations are possible regarding the mode: 1) If all values occur only once in a distribution, Excel will return #N/A.   2) If a variable has only one mode, Excel will return that value.   3) If a variable has more than one mode, however, Excel will still return only one value. The value that is returned will be the one associated with the modal value that occurs first in the data set. This caution also applies to the MODE function. To check the accuracy of the mode, I suggest that you create a frequency distribution.

- **Standard Deviation**. Unbiased estimate of the population standard deviation found by applying the formula

$$S = \sqrt{\frac{\sum (X - \overline{X})^2}{n-1}}$$

For the TV ownership distribution, S=2.5796.

- **Sample Variance** Standard deviation squared (S^2). For the TV distribution, S^2=6.6545.

- **Kurtosis**. A numerical index that describes a distribution with respect to its flatness or peakedness as compared to a normal distribution. A negative value characterizes a relatively flat distribution, whereas a positive value characterizes a relatively peaked distribution. Note that a distribution must contain at least four observations in order for kurtosis to be calculated. Kurtosis for our sample problem is 3.2128, indicating that the TV distribution is somewhat peaked as compared to a normal distribution.

- **Skewness**. A numerical index that characterizes the asymmetry of a distribution. Negative skew indicates that the longer tail extends in the direction of low values in the distribution. Positive skew, in contrast, indicates that the longer tail extends in the direction of the high values in the distribution.   Note that skewness cannot be computed when there are fewer than three observations in the distribution. The positive value of 1.6029 for our sample research problem indicates that the distribution of TV ownership is positively skewed. The positive skew is primarily due to the nine TVs owned by one of the survey respondents.

- **Range**. The minimum value in a distribution subtracted from the maximum value. The range for our distribution is found by subtracting 0 from 9, producing a range equal to 9.

- **Minimum**. The lowest value occurring in the distribution. The lowest number of TVs owned by customers in our sample is 0.

- **Maximum**. The highest value occurring in the distribution. The highest number of TVs owned by a customer in our sample is 9.

- **Sum**. Sum of the values in the distribution. For our sample research problem, ΣX=29.

- **Count**. Number of observations in the distribution. For our sample research problem, n=11.

- **Largest(2)**. Second largest value in the distribution. When the data are arranged in descending order, the second largest value is the number that occurs right after the number in first position. The second largest number of TVs owned by customers in our sample is 5.

- **Smallest(2)**. Second smallest value in the distribution. When the data are arranged in ascending order, the second smallest value is the number that occurs immediately after the number in first position. The second smallest number of TVs owned by customers in our sample is 0.

- **Confidence Level (95.0%)**. Amount of error subtracted from and added to the sample mean when constructing a confidence interval for the population mean. To calculate the confidence level using the Descriptive Statistics Analysis Tool, it is assumed that the population variance is not known and is estimated by the sample variance. The confidence interval takes the general form shown below.

$$\overline{X} \pm \left(t_{Crit}\right)\left(s_{\overline{X}}\right)$$

In this formula, $\overline{X}$ serves as an estimate of the population mean, and the error component is calculated by multiplying the standard error of the mean times the two-tailed critical value of t for the selected value of alpha.   The t critical value (2.2281) for our sample research problem is found in the t distribution with n−1=10 degrees of freedom and alpha equal to .05. The standard error is equal to .7778. The product of 2.2281 and 0.7778 is 1.7330. The 95% confidence interval, therefore, is:

$(2.6364-1.7330) \le \mu \le 2.6364+1.7330$

$.9034 \le \mu \le 4.3694$

Steps to Follow to Obtain the Population Variance and Population Standard Deviation

The variance and standard deviation in the Descriptive Statistics output are both calculated using formulas for unbiased population estimates. The denominator of the variance estimate formula is n − 1. If you, instead, want the population variance calculated using the formula, $\sigma^2 = \dfrac{\sum(X-\mu)^2}{N}$, it is very simple to use an Excel function to obtain this value.

1. Start with the output obtained in the previous section.

	A	B	C	D	E
1	TV	MICROWAVE		TV	
2	3	1			
3	0	1		Mean	2.636364
4	9	2		Standard Error	0.777791
5	2	1		Median	2
6	3	3		Mode	3
7	1	0		Standard Deviation	2.579641
8	2	1		Sample Variance	6.654545
9	0	2		Kurtosis	3.212809
10	3	1		Skewness	1.602855
11	5	1		Range	9
12	1	1		Minimum	0
13				Maximum	9
14				Sum	29
15				Count	11
16				Largest(2)	5
17				Smallest(2)	0
18				Confidence Level(95.0%)	1.733026

2. Click in cell **D7** and change the label to **Population Standard Dev**. Then click in cell **E7** where the population standard deviation will be placed.

	A	B	C	D	E
1	TV	MICROWAVE		TV	
2	3	1			
3	0	1		Mean	2.636364
4	9	2		Standard Error	0.777791
5	2	1		Median	2
6	3	3		Mode	3
7	1	0		Population Standard Dev	2.579641
8	2	1		Sample Variance	6.654545

3. Click **Insert** in the Menu Bar and select **Function**.

4. In the Insert Function dialog box, select the **Statistical** category, select the **STDEVP** function, and click **OK**.

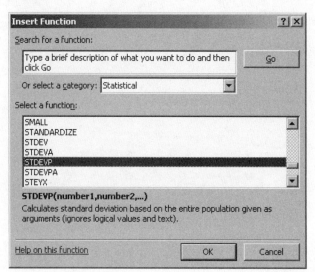

5. Click in the **Number 1** window. Then, in the worksheet, click and drag over the range A1:A12 to enter the TV data range. Click **OK**.

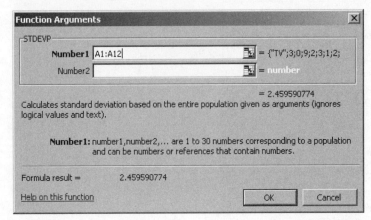

6. The function returns a population standard deviation of 2.4596. Click in cell **D8** and change the label to **Population Variance**. Then click in **E8** where the population variance will be placed.

	A	B	C	D	E
1	TV	MICROWAVE		*TV*	
2	3	1			
3	0	1		Mean	2.636364
4	9	2		Standard Error	0.777791
5	2	1		Median	2
6	3	3		Mode	3
7	1	0		Population Standard Dev	2.459591
8	2	1		Population Variance	6.654545
9	0	2		Kurtosis	3.212809
10	3	1		Skewness	1.602855
11	5	1		Range	9
12	1	1		Minimum	0
13				Maximum	9
14				Sum	29
15				Count	11
16				Largest(2)	5
17				Smallest(2)	0
18				Confidence Level(95.0%)	1.733026

7. Click **Insert** in the Menu Bar and select **Function**.

8. In the Insert Function dialog box, select the **Statistical** category, select the **VARP** function, and click **OK**.

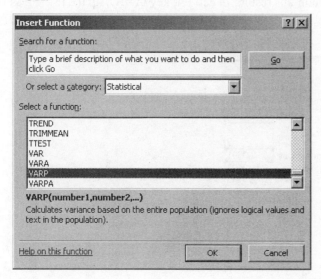

9. Click in the **Number 1** window. Then, in the worksheet, click and drag over the range A1:A12 to enter the TV data range. Click **OK**.

10. The function returns a population variance of 6.0496. Click in cell **E4** to change the value of the Standard Error.

	A	B	C	D	E
1	TV	MICROWAVE		*TV*	
2	3	1			
3	0	1		Mean	2.636364
4	9	2		Standard Error	0.777791
5	2	1		Median	2
6	3	3		Mode	3
7	1	0		Population Standard Dev	2.459591
8	2	1		Population Variance	6.049587
9	0	2		Kurtosis	3.212809
10	3	1		Skewness	1.602855
11	5	1		Range	9
12	1	1		Minimum	0
13				Maximum	9
14				Sum	29
15				Count	11
16				Largest(2)	5
17				Smallest(2)	0
18				Confidence Level(95.0%)	1.733026

11. When the population standard deviation is known, the standard error is calculated using the formula

$\sigma_{\bar{X}} = \dfrac{\sigma}{\sqrt{N}}$. We will key in a formula to obtain this result. In cell E4, key in **=E7/sqrt(E15)**. Press

[Enter].

	A	B	C	D	E
1	TV	MICROWAVE		*TV*	
2	3	1			
3	0	1		Mean	2.636364
4	9	2			=E7/sqrt(E15)
5	2	1		Median	2
6	3	3		Mode	3
7	1	0		Population Standard Dev	2.459591
8	2	1		Population Variance	6.049587
9	0	2		Kurtosis	3.212809
10	3	1		Skewness	1.602855
11	5	1		Range	9
12	1	1		Minimum	0
13				Maximum	9
14				Sum	29
15				Count	11
16				Largest(2)	5
17				Smallest(2)	0
18				Confidence Level(95.0%)	1.733026

12. The standard error is equal to 0.7416. Click in cell **E18** where the 95% confidence level will be placed.

	A	B	C	D	E
1	TV	MICROWAVE		*TV*	
2	3	1			
3	0	1		Mean	2.636364
4	9	2		Standard Error	0.741595
5	2	1		Median	2
6	3	3		Mode	3
7	1	0		Population Standard Dev	2.459591
8	2	1		Population Variance	6.049587
9	0	2		Kurtosis	3.212809
10	3	1		Skewness	1.602855
11	5	1		Range	9
12	1	1		Minimum	0
13				Maximum	9
14				Sum	29
15				Count	11
16				Largest(2)	5
17				Smallest(2)	0
18				Confidence Level(95.0%)	1.733026

13. The 95% confidence level is equal to the product of the standard error times the critical Z for a two-tailed test. For this problem, the standard error is found in cell E4 and the critical Z is 1.96. We will use a formula to carry out the calculations. In cell E18, key in =**E4*1.96**. Press [**Enter**].

	A	B	C	D	E
1	TV	MICROWAVE		TV	
2	3	1			
3	0	1		Mean	2.636364
4	9	2		Standard Error	0.741595
5	2	1		Median	2
6	3	3		Mode	3
7	1	0		Population Standard Dev	2.459591
8	2	1		Population Variance	6.049587
9	0	2		Kurtosis	3.212809
10	3	1		Skewness	1.602855
11	5	1		Range	9
12	1	1		Minimum	0
13				Maximum	9
14				Sum	29
15				Count	11
16				Largest(2)	5
17				Smallest(2)	0
18				Confidence Level(95.0%)	=E4*1.96

The 95% confidence level is equal to 1.4535. When you want population values instead of population estimates based on a sample, you need to change the standard error, the population standard deviation, the population variance, and the confidence level. The completed worksheet with these population values is shown below.

	A	B	C	D	E
1	TV	MICROWAVE		TV	
2	3	1			
3	0	1		Mean	2.636364
4	9	2		Standard Error	0.741595
5	2	1		Median	2
6	3	3		Mode	3
7	1	0		Population Standard Dev	2.459591
8	2	1		Population Variance	6.049587
9	0	2		Kurtosis	3.212809
10	3	1		Skewness	1.602855
11	5	1		Range	9
12	1	1		Minimum	0
13				Maximum	9
14				Sum	29
15				Count	11
16				Largest(2)	5
17				Smallest(2)	0
18				Confidence Level(95.0%)	1.453525

Steps to Follow to Obtain Descriptive Statistics for Two or More Variables

In order to use the Descriptive Statistics Analysis Tool to produce descriptive statistics for two or more variables simultaneously, the variables must be located in either adjacent columns or adjacent rows in the worksheet. In this section, I provide a brief explanation of how to obtain descriptive statistics for two variables. If a more complete description is required, you might want to refer back to the steps presented in the first section of this chapter.

1. If you have not already done so, enter the data in an Excel worksheet as shown below.

	A	B
1	TV	MICROWAVE
2	3	1
3	0	1
4	9	2
5	2	1
6	3	3
7	1	0
8	2	1
9	0	2
10	3	1
11	5	1
12	1	1

2. Click **Tools** in the Menu Bar and select **Data Analysis**.

If Data Analysis does not appear as a choice in the Tools menu, you will need to load the Microsoft Excel ToolPak add-in. Follow the procedure on page 7.

3. In the Data Analysis dialog box, select **Descriptive Statistics** click **OK**.

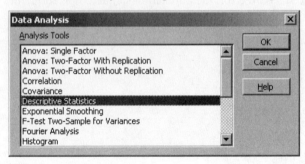

4. Complete the Descriptive Statistics dialog box as shown below. The entries are nearly the same as in the previous section except that the input range includes the worksheet location of both TV and MICROWAVE, **$A$1:$B$12**.

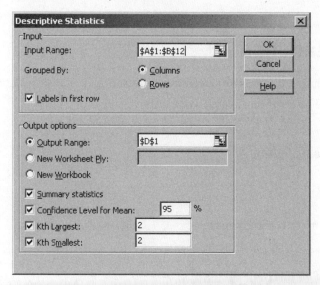

5. Click **OK** and you will be asked if you want to overwrite existing data. In this case, you do. Click **OK**.

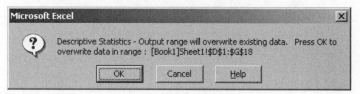

Adjust the column width as necessary to display the labels. For information regarding the interpretation of the output, see **Interpreting the Output** on pages 86 to 88.

	A	B	C	D	E	F	G
1	TV	MICROWAVE		*TV*		*MICROWAVE*	
2	3	1					
3	0	1		Mean	2.636363636	Mean	1.272727
4	9	2		Standard Error	0.777790896	Standard Error	0.237062
5	2	1		Median	2	Median	1
6	3	3		Mode	3	Mode	1
7	1	0		Standard Deviation	2.579640567	Standard Deviation	0.786245
8	2	1		Sample Variance	6.654545455	Sample Variance	0.618182
9	0	2		Kurtosis	3.212808683	Kurtosis	1.649366
10	3	1		Skewness	1.60285529	Skewness	0.935197
11	5	1		Range	9	Range	3
12	1	1		Minimum	0	Minimum	0
13				Maximum	9	Maximum	3
14				Sum	29	Sum	14
15				Count	11	Count	11
16				Largest(2)	5	Largest(2)	2
17				Smallest(2)	0	Smallest(2)	1
18				Confidence Level(95.0%)	1.733026106	Confidence Level(95.0%)	0.528207

Missing Values

Missing values should be indicated by blank cells. The count will reflect the number of nonblank cells in the range for each variable included in the input range.

Functions: Descriptive Statistics

We can use Excel's functions to obtain the same output that was produced by the Descriptive Statistics Data Analysis Tool, but it requires a little more work. The functions related to descriptive statistics are listed below.

Function Name	Value Returned by the Function
AVEDEV	Average deviation
AVERAGE	Mean
COUNT	Number of observations in a specified range
KURT	Kurtosis
LARGE	Kth observation in a range if values are arranged in descending order
MAX	Maximum
MEDIAN	Median
MIN	Minimum
MODE	Mode
QUARTILE	First, second, third, and fourth quartile
SKEW	Skewness
SMALL	Kth observation in a range if observations are arranged in ascending order
STDEV	Standard deviation calculated using the unbiased estimate formula
STDEVP	Standard deviation calculated using the population formula
VAR	Variance calculated using the unbiased estimate formula
VARP	Variance calculated using the population formula

Sample Research Problem

To illustrate the use of Excel's functions for obtaining descriptive statistics, I will use the same set of data that was utilized at the beginning of this chapter to illustrate the Descriptive Statistics Analysis Tool.

Steps to Follow to Obtain Descriptive Statistics for One Variable

I have chosen the TV variable to illustrate how to use functions to obtain descriptive statistics for one variable. I will provide detailed instructions for only a couple of the statistics because the procedure is quite repetitive. Once you have the desired descriptive statistics for one variable, obtaining them for additional variables is a relatively simple process, as long as the variables are located in adjacent columns of the worksheet.

1. If you have not already done so, enter the data shown below in an Excel worksheet.

	A	B
1	TV	MICROWAVE
2	3	1
3	0	1
4	9	2
5	2	1
6	3	3
7	1	0
8	2	1
9	0	2
10	3	1
11	5	1
12	1	1

2. **AVEDEV**. Let's begin with the average deviation. The average deviation is found by first calculating the absolute value of the deviation score for each observation $\left| X - \overline{X} \right|$ and then calculating the average.

First let's provide a label for the output. Activate **A14** and key in **AVEDEV**.

Labels are not necessary, but they help to make the output more informative, especially if you refer to it a few days after creating it.

11	5	1
12	1	1
13		
14	AVEDEV	

3. Activate cell **B14** where the average deviation will be placed.

4. Click **Insert** in the Menu Bar and select **Function**.

5. In the Insert Function dialog box, select the **Statistical** category and the **AVEDEV** function. Click **OK**.

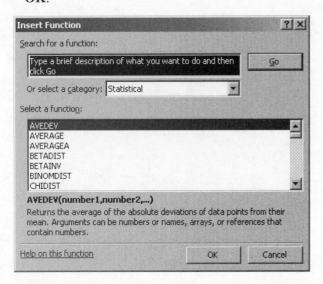

6. Click in the **Number 1** window. Then click in cell **A2** of the worksheet where the TV data begin and drag to cell **A12** where the data end. If you prefer, you can manually enter **A2:A12** in the Number 1 window.

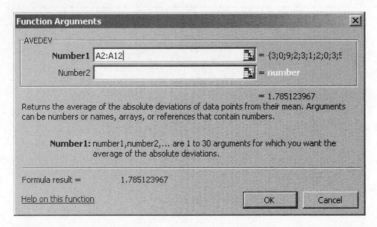

7. Click **OK**. The AVEDEV function returns a value of 1.7851 and places it in cell B14 of the worksheet.

11	5	1
12	1	1
13		
14	AVEDEV	1.785123967

8. **AVERAGE**. Next let's find the average (or mean) by using the AVERAGE function. Activate cell **A15** and enter the label **AVERAGE**.

11	5	1
12	1	1
13		
14	AVEDEV	1.785123967
15	AVERAGE	

9. Activate cell **B15** where the average will be placed.

10. Click **Insert** in the Menu Bar and select **Function**.

11. In the Insert Function dialog box, select the **Statistical** category and the **AVERAGE** function. Click **OK**.

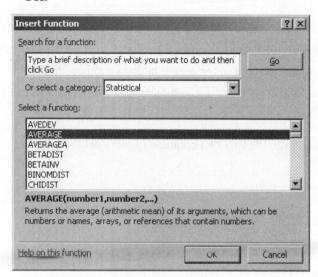

12. Click in the **Number 1** window. Then click on cell **A2** where the TV data begin and drag to cell **A12** where the TV data end.   If you prefer, you can manually enter **A2:A12** in the Number 1 window.

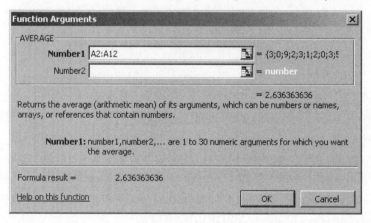

13. Click **OK** and a value of 2.6364 is returned and placed in cell B14 of the worksheet.

10	3	1
11	5	1
12	1	1
13		
14	AVEDEV	1.785123967
15	AVERAGE	2.636363636

As you can see, regardless of the function that you are using, the steps that you will follow are very similar.
If you run into any difficulty, you can read the description provided at the bottom of the Paste Function dialog

box and at the bottom of the dialog box for a specific function. If you need additional information, just click the question mark in the lower left corner of the function's dialog box.

Steps to Follow to Obtain Descriptive Statistics for Two or More Variables

As I said earlier, once you have descriptive statistics for one variable, obtaining them for additional variables is a very simple process, as long as the variables appear in adjacent columns (or rows) in the worksheet. You essentially just copy the functions and formulas from one column in the output table to another. In the previous section, we found the average deviation and the average for TV and placed the values in column B. Now let's obtain those two values for MICROWAVE.

1. When we completed the final step in the last section, the first two columns of our worksheet looked like those shown below.

	A	B
1	TV	MICROWAVE
2	3	1
3	0	1
4	9	2
5	2	1
6	3	3
7	1	0
8	2	1
9	0	2
10	3	1
11	5	1
12	1	1
13		
14	AVEDEV	1.785123967
15	AVERAGE	2.636363636

Let's continue by finding the average deviation for MICROWAVE and placing it in C14. Click in cell **B14** where the average deviation for TV is located. If you look up in the formula bar, you will see the contents of cell B14. Those contents are =AVEDEV(A2:A12). AVEDEV is the function name and A2:A12 is the data range. If you copy the contents of B14 to C14, the column referent will change from A to B, but everything else will stay the same. That's because the cell addresses are relative rather than absolute. So, copy B14 to C14. You should now see a value of 0.5785 in cell C14. Click on C14 to check its contents. In the formula bar you will see =AVEDEV(B2:B12).

14	AVEDEV	1.785123967	0.578512
15	AVERAGE	2.636363636	

2. Next, let's find the average for MICROWAVE. Click in cell **B15** where the average for TV is located. Look in the formula bar and you will see =AVERAGE(A2:A12). Copy the contents of B15 to C15. You should now see a value of 1.2727 in cell C15. Click on C15 to check its contents. In the formula bar you will see =AVERAGE(B2:B12).

14	AVEDEV	1.785123967	0.578512
15	AVERAGE	2.636363636	1.272727

Pivot Table: Descriptive Statistics

The Pivot Table is a tool that is very useful for summarizing the variables contained in a database. It is located in the Data menu. Be sure to provide a variable label at the top of each column in the worksheet, because labels are necessary when constructing a Pivot Table.

Sample Research Problem

This research problem is the same one that is presented in Chapter 4. Students enrolled in an introductory world religions course provided information that the instructor used for assigning students to discussion and project groups. The data set is shown below.

	A	B	C
1	SEX	AGE	RELIGION
2	M	19	Catholic
3	M	20	Jewish
4	F	22	None
5	M	23	Muslim
6	F	31	Catholic
7	F	22	Protestant
8	F	10	None
9	M	20	Jewish
10	M	19	Buddhist
11	F	20	Protestant
12	F	20	Protestant
13	M	19	Catholic
14	F	21	Protestant
15	M	19	Jewish
16	F	19	Catholic
17	F	20	Protestant

Steps to Follow to Set up a Pivot Table

Using the Pivot Table, we will be able to obtain the following values: sum, count, average, maximum, minimum, sample standard deviation, population standard deviation, sample variance, and population variance. Let's say that the instructor of the world religions course wanted summary information regarding AGE for the entire class and for each major separately.

1. If you have not already done so, enter the student data shown above in an Excel worksheet

2. Activate any cell in the data range (A1 to F17).

If a cell in the data range is active before Pivot Table is selected, then Excel will automatically enter the data range in the dialog box.

---

3. Click **Data** in the Menu Bar and select **Pivot Table and Pivot Chart Report**.

4. **Pivot Table and Pivot Chart Wizard-Step 1 of 3**. The data are located in an Excel worksheet, so select **Microsoft Excel List or Database**. You want to create a Pivot Table report, so select **Pivot Table**. Click **Next>**.

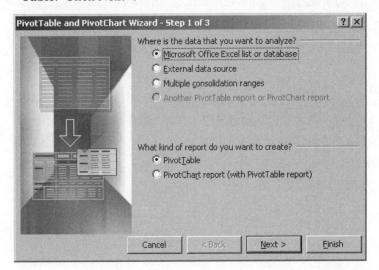

5. **Pivot Table and Pivot Chart Wizard-Step 2 of 3**. Enter the range of the data (A1 to C17) if it has not been automatically entered by Excel. If it has been, check to make sure that it is correct. If it has not been entered, click in the **Range** window. Then click in cell **A1** of the worksheet and drag to cell **C17**. If you prefer, you can manually enter **A1:C17** in the Range window. Click **Next>**.

> *You might see a message that starts out "Your new report will use less memory if you base it on your existing report . . . " Unless memory is a problem, I usually click **No**.*

6. **Pivot Table and Pivot Chart Wizard-Step 3 of 3**. Let's put the report in a new worksheet. Click in the button to the left of **New worksheet** to place a black dot there. Click **Finish**.

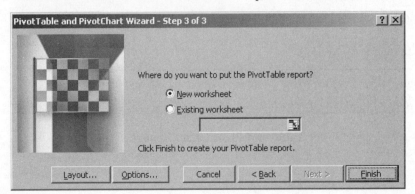

7. **Count**. We will first display the count (number of observations) for AGE broken down by SEX. Drag the **SEX** field button from the Pivot Table Field List to the area labeled **Drop Column Fields Here**.

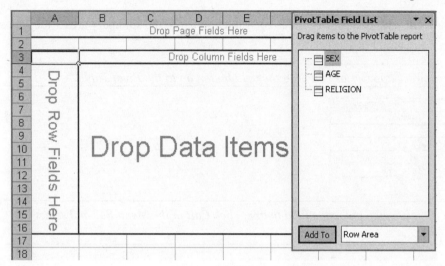

8. Drag the **AGE** field button to the area labeled **Drop Data Items Here**. The default summary measure for a quantitative variable is sum. So you will see Sum of AGE in cell A3. We want to change it to Count.

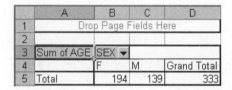

9. Double-click in cell A3 on **Sum of AGE**. In the Pivot Table Field dialog box, select **Count**.

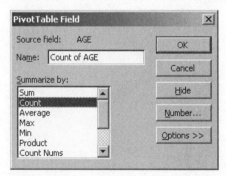

10. Click **OK** in the Pivot Table Field dialog box. You will now see counts rather than sums in the cells of the table.

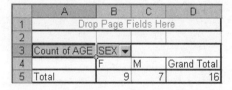

11. In order to select more summary measures, the Pivot Table Field List must be displayed. If it isn't, just click in a cell of the Pivot Table. I clicked in cell B5.

12. **Mean**. Drag the **AGE** field button from the Pivot Table toolbar to the summary area—the area shown below that has a fuzzy gray border (B5:D5).

The fuzzy gray border appears as you drag the field button on to the Pivot Table.

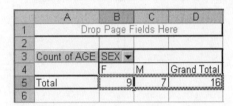

*If you make a mistake when placing a field button, click **Edit** in the Menu Bar and select **Undo Pivot**.*

The default summary measure, sum of GPA, is displayed.

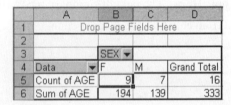

13. You need to change Sum of GPA to Average. Right-click on **Sum of GPA** in cell A6. Select **Field Settings** from the drop down menu that appears.

14. In the Pivot Table Field dialog box, select **Average** in the Summarize by window.

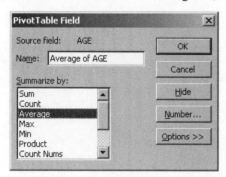

15. Click **OK**. You now see Average of AGE in A6 and numerical averages displayed in the other cells of row 6.

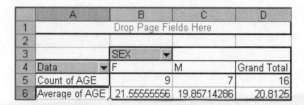

16. **Changing the name**. Excel gives you the option of changing the names of the summary measures. Let's say that we prefer to use the expression *mean* rather than *average*. To make this change, first right-click on **Average of GPA** in cell A6. Select **Field Settings** from the drop down menu that appears.

17. Find the Name window in the top left of the Pivot Table Field dialog box. Change the entry in the window from Average of GPA to **Mean of GPA**. Click **OK**.

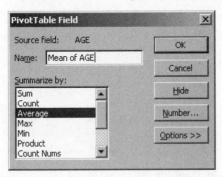

Mean of GPA is now displayed in cell A6 of the Pivot Table.

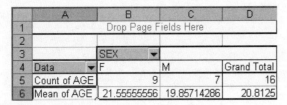

	A	B	C	D
1		Drop Page Fields Here		
2				
3		SEX ▼		
4	Data ▼	F	M	Grand Total
5	Count of AGE	9	7	16
6	Mean of AGE	21.55555556	19.85714286	20.8125

18. **Changing the number of decimals**. Mean GPA is a calculated value that could have several decimal places. Let's say that we want the value displayed with two decimal places. To make this change, first right-click on **Mean of GPA** in cell A6. Select **Field Settings** from the drop down menu that appears.

19. Click the **Number** button on the right side of the Pivot Field dialog box.

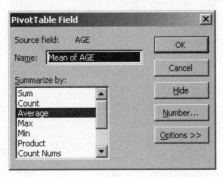

20. In the Format Cells dialog box shown below, select the **Number** Category and set Decimal places equal to **2**. Click **OK** in the Format Cells dialog box. Click **OK** in the Pivot Table Field dialog box.

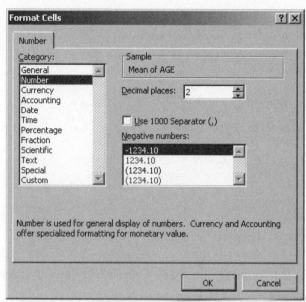

All means will now be displayed with 2 decimal places.

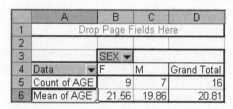

21. To add another summary measure, the Pivot Table Field List must be displayed. If it is not, click in a cell of the Pivot Table. I clicked in cell B5.

22.   **Standard deviation**. Let's add the standard deviation to the Pivot Table. Drag **AGE** to the Pivot Table—the area shown below that has a fuzzy gray border (B5:E6).

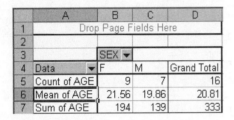

The default summary measure, Sum of AGE, is now displayed in row 7.

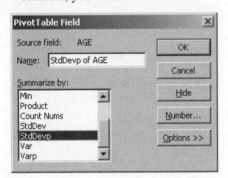

23.   To change this to the standard deviation, right-click **Sum of GPA** in A7. Select **Field Settings** from the drop down menu that appears.

24.   Select **StdDevp** in the Summarize by window. *StdDevp* stands for the standard deviation of a population. The population formula is used rather than the sample formula. If you want the sample formula, you must select *StdDev*.

PivotTable Field		
Source field: AGE		OK
Name: StdDevp of AGE		Cancel
Summarize by:		Hide
Min		
Product		Number...
Count Nums		
StdDev		Options >>
StdDevp		
Var		
Varp		

25.   Click **OK** and StdDevp of GPA appears in row 7 the Pivot Table.

	A	B	C	D
1		Drop Page Fields Here		
2				
3		SEX		
4	Data	F	M	Grand Total
5	Count of AGE	9	7	16
6	Mean of AGE	21.56	19.86	20.81
7	StdDevp of AGE	3.499559055	1.355261854	2.898679656

26. **Minimum**. Let's add minimum to the Pivot Table. Drag the **AGE** Pivot Table Field List to the Pivot Table, B5:D7. Sum of GPA will now be displayed in row 8.

	A	B	C	D
1		Drop Page Fields Here		
2				
3		SEX ▼		
4	Data ▼	F	M	Grand Total
5	Count of AGE	9	7	16
6	Mean of AGE	21.56	19.86	20.81
7	StdDevp of AGE	3.499559055	1.355261854	2.898679656
8	Sum of AGE	194	139	333

27. To change Sum of AGE to Min of AGE, right-click on **Sum of AGE**. Select **Field Settings** from the drop down menu that appears.

28. Select **Min** in the Summarize by window.

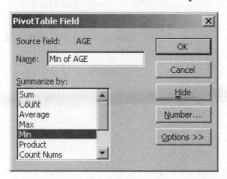

29. Click **OK**. Min of AGE now appears in row 8 of the Pivot Table.

	A	B	C	D
1				
2				
3		SEX ▼		
4	Data ▼	F	M	Grand Total
5	Count of AGE	9	7	16
6	Mean of AGE	21.56	19.86	20.81
7	StdDevp of AGE	3.499559055	1.355261854	2.898679656
8	Min of AGE	19	19	19

As you can see, Pivot Tables are fairly easy to construct. In addition, they can be customized to meet the user's needs. Several summary measures may be displayed in a single Pivot Table, several formats are available (e.g., number, currency, date), and grand totals are optional.

Removing Summary Measures from a Pivot Table

When you are working with a Pivot Table, you might decide that some of the summary measures that you included are not really necessary. Removing them is a simple process. For example, let's say that we want to remove Min of GPA from the Pivot Table created in the previous section.

1. Click on the arrow to the right of **Data** in cell A4 of the table.

2.   A listing of the summary measures included in the Pivot Table will appear.

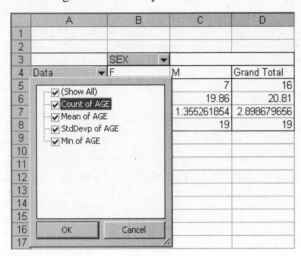

3.   Click in the box to the left of any summary measure that you would like to remove. For this example, we will remove **Min of AGE**.

4.   Click **OK**. Min of AGE has now been removed from the Pivot Table.

	A	B	C	D
1				
2				
3		SEX ▼		
4	Data ▼	F	M	Grand Total
5	Count of AGE	9	7	16
6	Mean of AGE	21.56	19.86	20.81
7	StdDevp of AGE	3.499559055	1.355261854	2.898679656

Chapter 6

Probability Distributions

Excel provides functions that can be utilized for calculating the probabilities associated with a number of different distributions. Because the required hand calculations are often quite time consuming, you will appreciate the speed with which Excel can compute the probabilities for you. In this chapter, I first present discrete probability distributions and then continuous probability distributions.

Discrete Probability Distributions

Discrete variables can assume, on any single trial, a countable set of values. A variable is considered a discrete random variable if the probability of all possible values can be specified. Introductory statistics textbooks often use coin tossing and random selection of research participants from a subject pool to illustrate probabilities of discrete random variables. I will describe how Excel's functions can be used to answer questions regarding three discrete probability distributions: binomial, hypergeometric, and Poisson.

Binomial Distribution

Binomial variables take on only two values. One of these values is typically designated as a "success" and the other as a "failure." The probability of a "success" is symbolized as p, and the probability of a "failure" as q. Because "success" and "failure" are the only possible outcomes, the sum of their probabilities is 1.

We will use the example of tossing a fair coin five times to illustrate how to utilize Excel's functions for computing probabilities for a binomial distribution. We will consider "heads" a success and "tails" a failure. Because we are tossing a fair coin, the probability of a success on a single trial is .5. The outcomes are independent, because the probability of tossing heads on one trial does not change the probability of tossing heads on a subsequent trial.

The general form of the probability function for the binomial distribution is given by

$$P(X = x) = \binom{N}{x} p^x q^{N-x}$$

where P = probability.

X = the event of interest. For our example, the event of interest is tossing heads.

x = the specific number of times the event of interest occurs. For our example of five tosses, x can take on values from 0 to 5.

N = the number of tosses (or trials). For our example, N = 5 tosses.

p = the probability of success. For our example, success is tossing heads. Because the coin is fair, p = .5.

q = the probability of failure. For our example, failure is tossing tails. Because (p + q) = 1 and p = .5, q must also be equal to .5.

P(X = x), therefore, stands for the probability that the number of tosses that result in heads is equal to a specific number (0, 1, 2, 3, 4, or 5).

We will display the probabilities for this binomial distribution in an Excel worksheet.

Binominal Probabilities

1. Complete an Excel worksheet with entries as shown below. Include the labels, **X** and **P(X=x)**, in cells A1 and B1, respectively.

	A	B
1	X	P(X=x)
2	0	
3	1	
4	2	
5	3	
6	4	
7	5	

2. **P(X = 0)**. This probability will be placed in cell B2 of the output table, so activate cell **B2**.

3. Click **Insert** in the Menu bar and select **Function**.

4. In the Insert Function dialog box, select the **Statistical** function category and the **BINOMDIST** function. Click **OK**.

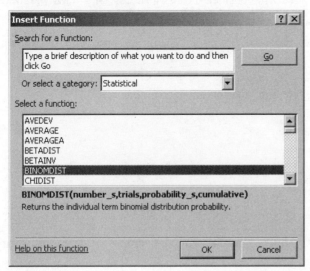

5. Complete the BINOMDIST dialog box as shown below. Details regarding the entries are given immediately after the dialog box.

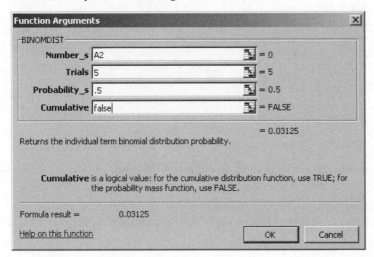

- **Number_s**. The number of successes is equal to 0. We could enter 0 here, but I recommend that we enter a cell address, because that will enable us to copy the function to the other cells in the output table. That will save us some work later on. So click in the **Number_s** window and enter **A2**. (A2 is the cell location of 0.)

- **Trials**. The total number of trials (tosses) is equal to 5. Click in the **Trials** window and enter **5**.

- **Probability_s**. The probability of a success, or tossing heads with a fair coin, is .5. Click in the **Probability_s** window and enter **.5**.

- **Cumulative**. We do not want cumulative probabilities in column B of the worksheet, so enter **false**.

6. Click **OK** and a value of .03125 is returned. This is the probability of tossing a fair coin five times and having none of the tosses result in a head (0 successes).

	A	B
1	X	P(X=x)
2	0	0.03125
3	1	
4	2	
5	3	
6	4	
7	5	

7. **Probabilities for 1 to 5 successes**. To obtain the remaining probabilities, we can copy the **BINOMDIST** function that was entered in cell B2 to cells B3 through B7. Activate cell **B2**. Look in the Formula Bar to make sure that the cell entry is the BINOMDIST function, BINOMDIST (A2, 5, 0.5, false).

8. Use the fill handle to copy the BINOMDIST function. To do this, move the pointer to the lower right corner of cell B2. It will turn into a black plus sign. Press the left mouse key and hold it down while

you drag to cell B7. Release the mouse key and the probabilities for 1 to 5 successes will be returned. Check the accuracy of your output by referring to the worksheet shown below.

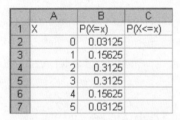

	A	B
1	X	P(X=x)
2	0	0.03125
3	1	0.15625
4	2	0.3125
5	3	0.3125
6	4	0.15625
7	5	0.03125

Cumulative Binomial Probabilities

The BINOMDIST function can also be used to obtain cumulative probabilities. We will use the same output table that we created in the last section.

1. **Label**. In cell **C1** of the output table, enter the label **P(X<=x)**. This stands for the probability that X (the event of interest) is less than or equal to x (specific values ranging from 0 to 5).

	A	B	C
1	X	P(X=x)	P(X<=x)
2	0	0.03125	
3	1	0.15625	
4	2	0.3125	
5	3	0.3125	
6	4	0.15625	
7	5	0.03125	

2. **P(X<=0)**. The probability that X is less than or equal to 0 is a cumulative probability. Activate cell **C2** where this cumulative probability value will be placed.

3. Click **Insert** in the Menu Bar and select **Function**.

4. In the Insert Function dialog box, select the **Statistical** category and select the **BINOMDIST** function. Click **OK**.

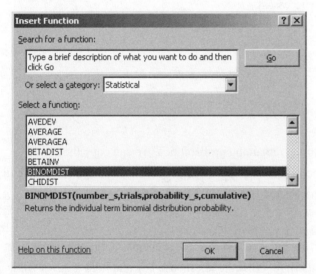

5. Complete the BINOMDIST dialog box as shown below. Detailed information regarding the entries is given immediately after the dialog box.

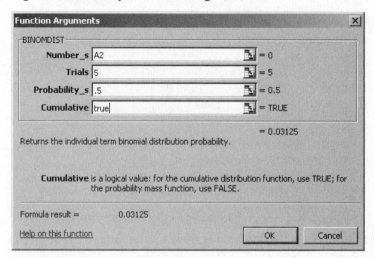

- **Number_s**. Click in the **Number_s** window and enter **A2**, the cell location of 0 (the number of successes).

- **Trials**. Click in the **Trials** window and enter **5**, the total number of trials (tosses).

- **Probability_s**. Click in the **Probability_s** window and enter **.5**, the probability of a success (heads) on a single trial.

- **Cumulative**. Click in the **Cumulative** window and enter **true**, because we want the cumulative probabilities in Column C of the output table.

6. Click **OK** and a value of .03125 is returned.

	A	B	C
1	X	P(X=x)	P(X<=x)
2	0	0.03125	0.03125
3	1	0.15625	
4	2	0.3125	
5	3	0.3125	
6	4	0.15625	
7	5	0.03125	

7. **Cumulative probabilities for 1 to 5 successes**. To obtain the remaining cumulative probabilities, we will copy the BINOMDIST function in cell C2 to cells C3 through C7. Activate cell **C2** where the cumulative probability associated with zero successes was placed.

Note that it is essential that a cell address (A2) is entered in the dialog box for Number_s in order for us to copy the function accurately.

---

8. Use the fill handle to copy the BINOMDIST function. To do this, move the pointer to the lower right corner of cell C2. It will turn into a black plus sign. Press the left mouse key and hold it down while you drag to cell C7. Release the mouse key and the cumulative probabilities for 1 to 5 successes will be returned. Check the accuracy of your output by referring to the worksheet shown below.

	A	B	C
1	X	P(X=x)	P(X<=x)
2	0	0.03125	0.03125
3	1	0.15625	0.1875
4	2	0.3125	0.5
5	3	0.3125	0.8125
6	4	0.15625	0.96875
7	5	0.03125	1

Interpreting the Output

For our sample problem, the probability of tossing a fair coin five times and obtaining zero heads is .03125. The probability of tossing the coin five times and obtaining one head is .15625, and so on. The most likely outcomes are two heads or three heads, each associated with a probability of .3125.

The cumulative probabilities displayed in the output table represent the likelihood of obtaining heads on x trials or less. For example, the probability of obtaining heads on three trials or less (i.e., three trials, two trials, one trial, or zero trials) is .8125. The probability of obtaining heads on five trials or less, of course, is equal to 1 because the coin is being tossed only five times.

Hypergeometric Distribution

The HYPGEOMDIST function returns probabilities associated with the hypergeometric distribution. The hypergeometric distribution is similar to the binomial distribution except we do not have independence. More specifically, the outcome on one trial changes the probability of an outcome on a subsequent trial. We will use an example of randomly sampling participants for a research project to illustrate how to use the HYPGEOMDIST function. Let's say that 15 volunteers signed up to participate in the research project but we only need three subjects. We know that nine of the volunteers are women and that six are men. If we randomly sample three volunteers, what is the probability that two will be women and one will be a man? In this situation, we are sampling without replacement, because after a person is selected we do not want to select that person again. Before selection begins, the probability of sampling a woman is 9/15, or .6. If the first draw is a man, then the probability of sampling a woman on the second draw has changed to 9/14, or .64, because only 14 persons remain in the subject pool.

The general form of the probability function for the hypergeometric distribution is given by

$$P(X=x)=\frac{\binom{M}{x}\binom{N-M}{n-x}}{\binom{N}{n}}$$

where   M = the number of successes in the population. The number of women in the population is 9.

x = the number of successes in the sample. We will declare selecting a woman a success.

N = the size of the population. For our example, N = 15 volunteers.

n = the size of the sample. We are drawing a sample of size n = 3.

We will display the probabilities for this research participant sampling problem in an Excel worksheet.

1. Complete the worksheet with entries as shown below. The X in this worksheet refers to selecting a woman as a research project participant. Because we are randomly sampling a total of three subjects, the specific values this event can take on are 0, 1, 2, and 3. P(X − x) refers to the probabilities associated with each of these specific outcomes.

	A	B
1	X	P(X=x)
2	0	
3	1	
4	2	
5	3	

2. **P(X = 0)**. The probability that the sample will contain zero women will be placed in cell B2 of the table, so activate cell **B2**.

3. Click **Insert** in the Menu Bar and select **Function**.

4. In the Insert Function dialog box, select the **Statistical** category and the **HYPGEOMDIST** function.
    Click **OK**.

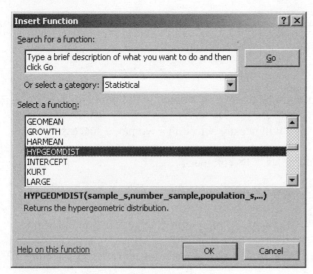

5. Complete the HYPGEOMDIST dialog box as shown below. Detailed instructions are given
    immediately following the dialog box.

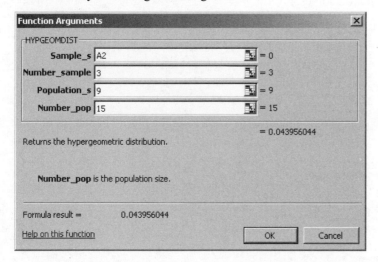

- **Sample_s**. The number of successes in the sample is equal to zero for our problem of finding
  $P(X = 0)$. Rather than entering 0 in this space, I suggest that you enter the cell address of 0 (A2) so
  that you will be able to copy the function to obtain the probabilities for x values 1, 2, and 3. So, click
  in the **Sample_s** window and enter **A2**.

- **Number_sample**. Click in the **Number_sample** window and enter **3**, the sample size.

- **Population_s**. Click in the **Population_s** window and enter **9**, the number of successes in the
  population.

- **Number_pop**. Click in the **Number_pop** window and enter **15**, the total number of volunteers in the population.

6. Click **OK** and a value of .0440 is returned. This is the probability of drawing a random sample of three research subjects and having none of the subjects be a woman (0 successes).

7. **Probabilities for 1 to 3 successes**. To obtain the remaining probabilities, we will be able to copy the HYPGEOMDIST function that was entered in cell B2 to cells B3 through B5. Recall that we can copy the function only because we entered a cell address for Sample_s in the HYPGEOMDIST dialog box when we requested the probability for zero successes. Activate cell **B2**.

8. Copy the function that appears in cell B2 to cells B3 through B5. Check the accuracy of your work by referring to the completed worksheet shown below.

	A	B
1	X	P(X=x)
2	0	0.043956
3	1	0.296703
4	2	0.474725
5	3	0.184615

Interpreting the Output

Given the parameters of our research participant selection problem, the probability that a sample of size n = 3 will contain zero women is .0440. The probability that a sample of size n = 3 will contain one woman is .2967, and so on. The most likely outcome is a sample with two women and one man, a combination that has a probability of .4747.

Poisson Distribution

The Poisson distribution is especially applicable in situations where the population is very large and the probability associated with a particular event is very small. The general expression for the probability function of the Poisson distribution is given by

$$P(X) = \frac{e^{-\lambda} \lambda^X}{X!}$$

where   e = a mathematical constant equal approximately to 2.7183,

λ = the expected number of successes and

X = the number of events.

To illustrate the use of the Poisson distribution, let's assume that the annual rate of mononucleosis among college students on campuses in the United States is 1 in 10,000 students, or .0001. On one particular campus with an enrollment of 15,000, ten students were diagnosed as having mononucleosis. What is the probability of ten such cases if the true rate is .0001? Should the college administrators be alarmed?

Poisson Probabilities

1. Complete a worksheet with entries as shown below. We will use this as an output table to display our work.

	A	B
1	X	P(X)
2	0	
3	1	
4	2	
5	3	
6	4	
7	5	
8	6	
9	7	
10	8	
11	9	
12	10	

2. **P(X = 0)**. Let's start with X = 0 and then use the copy command to obtain the probabilities associated with X values of 1 through 10. The probability that X = 0 will be displayed in cell B2, so activate cell **B2**.

3. Click **Insert** in the Menu Bar and select **Function**.

4. In the Insert Function dialog box, select the **Statistical** category and the **POISSON** function. Click **OK**.

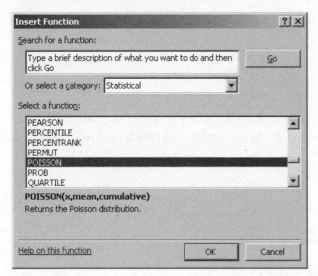

5. Complete the POISSON dialog box as shown below. Detailed instructions are given immediately following the dialog box.

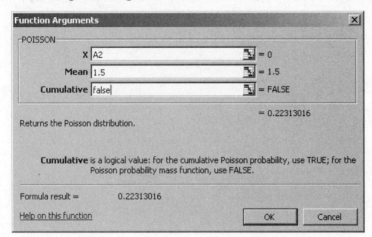

- **X** is the number of times the event of interest (cases of mononucleosis) occurs. We are computing the probability for X = 0 cases. Rather than entering a zero, however, we will enter A2, the cell location of zero in our output table. Click in the **X** window and enter **A2**.

- The **Mean** is the expected value assuming the true rate is .0001. To compute the expected value, we multiply the population size (15,000) times the true rate (.0001), which gives us 1.5. Click in the **Mean** window and enter **1.5**.

- The **Cumulative** window will contain either true or false, depending on whether or not we want cumulative probabilities. We do not want cumulative probabilities. So click in the **Cumulative** window and enter **false**.

6. Click **OK** and a value of .2231 is returned.

	A	B
1	X	P(X)
2	0	0.22313
3	1	
4	2	
5	3	
6	4	
7	5	
8	6	
9	7	
10	8	
11	9	
12	10	

7. **Probabilities for 1 to 10 cases**. To find the probabilities for 1 to 10 cases of mononucleosis on this particular college campus, we will copy the POISSON function that was entered in cell B2 of the output table to cells B3 through B12. (Note that X had to be specified by a cell address in the function arguments in order for us to copy it and obtain correct values.) Activate cell **B2**.

8. Copy the POISSON function in cell B2 to cells B3 through B12.   Check the accuracy of your work by referring to the completed worksheet shown below. You will see that the probabilities associated with 9 and 10 cases are displayed in scientific notation. This is because the probabilities are very small.

	A	B
1	X	P(X)
2	0	0.22313
3	1	0.334695
4	2	0.251021
5	3	0.125511
6	4	0.047067
7	5	0.01412
8	6	0.00353
9	7	0.000756
10	8	0.000142
11	9	2.36E-05
12	10	3.55E-06

Cumulative Poisson Probabilities

We can also use Excel's POISSON function to obtain cumulative probabilities. The cumulative probabilities represent the likelihood that X is less than or equal to a specified value. For example, what is the probability that the number of cases of mononucleosis on a campus with 15,000 students will be less than or equal to 5 if the true rate is .0001? We will use the same output table that was created in the previous section.

1. **Label**. In cell **C1** of the output table, enter the label **P(X<=x)**. This stands for the probability that X (cases of mononucleosis) is less than or equal to x (specific values ranging from 0 to 10).

	A	B	C
1	X	P(X)	P(X<=x)
2	0	0.22313	
3	1	0.334695	
4	2	0.251021	
5	3	0.125511	
6	4	0.047067	
7	5	0.01412	
8	6	0.00353	
9	7	0.000756	
10	8	0.000142	
11	9	2.36E-05	
12	10	3.55E-06	

2. Let's begin with P(X <= 0). Activate cell **C2** in the output table.

3. Click **Insert** in the Menu Bar and select **Function**.

4. In the Insert Function dialog box, select the **Statistical** category and the **POISSON** function. Click **OK**.

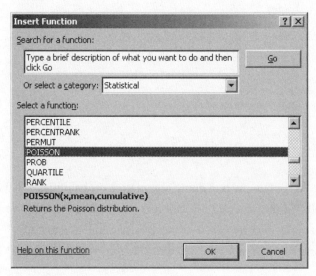

5. Complete the POISSON dialog box as shown below. Detailed information regarding the entries is given immediately following the dialog box.

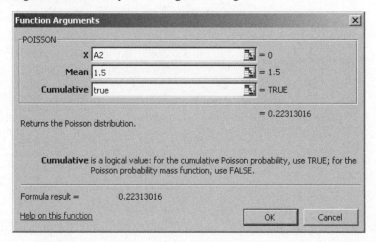

- **X**. Click in the **X** window and enter **A2**, the cell address for 0.

- **Mean**. Click in the **Mean** window and enter the expected value, **1.5**.

- **Cumulative**. Click in the **Cumulative** window and enter **true** to indicate that a cumulative probability is desired.

6. Click **OK** and a value of .2231 is returned.

	A	B	C
1	X	P(X)	P(X<=x)
2	0	0.22313	0.22313
3	1	0.334695	
4	2	0.251021	
5	3	0.125511	
6	4	0.047067	
7	5	0.01412	
8	6	0.00353	
9	7	0.000756	
10	8	0.000142	
11	9	2.36E-05	
12	10	3.55E-06	

7. **Cumulative probabilities for 1 to 10 cases**. To obtain the remaining cumulative probabilities, we will copy the POISSON function in cell C2 to cells C3 through C12. Recall that a cell address must have been entered for X in the POISSON function dialog box for X = 0 in order for the function to copy accurately. Activate cell **C2** where the cumulative probability associated with zero was placed.

8. Copy the cumulative POISSON function in cell C2 to cells C3 through C12. Check the accuracy by referring to the output table shown below.

	A	B	C
1	X	P(X)	P(X<=x)
2	0	0.22313	0.22313
3	1	0.334695	0.557825
4	2	0.251021	0.808847
5	3	0.125511	0.934358
6	4	0.047067	0.981424
7	5	0.01412	0.995544
8	6	0.00353	0.999074
9	7	0.000756	0.99983
10	8	0.000142	0.999972
11	9	2.36E-05	0.999996
12	10	3.55E-06	0.999999

Interpreting the Output

If the true rate of occurrence of mononucleosis is .0001, then, on a campus of 15,000 students, the probability of zero cases of mononucleosis is .2231. The most likely number of occurrences is 1, which is associated with a probability of .3347. The least likely number of occurrences (of the values we included in the output table) is 10, which is associated with a probability of 3.55E-06, or .00000355. Therefore, the campus administrators should be alarmed if as many as 10 cases are reported.

The probability of one case or less of mononucleosis on this campus is .5578. The probability of two cases or less jumps to .8088.

Continuous Probability Distributions

Continuous probability distributions play a major role in statistics because most of the quantitative variables analyzed by researchers are measured on a continuous scale. Furthermore, continuous probability distributions often provide very good approximations for discrete random variables. Unlike discrete variables, continuous variables can take on an infinite number of values between two points. Good examples of such variables are weight, distance, and time. I have included four continuous probability distributions in this section: normal, t, F, and chi-square. I devote most attention to the normal distribution, and only briefly explain how to use Excel's functions in place of textbook tables for the t, F, and chi-square distributions.

Normal Distribution

Excel provides several functions related to the normal distribution.

NORMDIST returns the normal cumulative distribution for specified X, mean, and standard deviation.

NORMINV returns the inverse of the normal cumulative distribution for specified probability, mean, and standard deviation.

NORMSDIST returns the standard normal cumulative distribution for a specified Z value.

NORMSINV returns the inverse of the standard normal cumulative distribution for a specified Z value.

STANDARDIZE returns a standardized score value for specified X, mean, and standard deviation.

In the examples that follow, I will illustrate how to use NORMSDIST, NORMSINV, and STANDARDIZE.

Calculating Standard Scores Using the STANDARDIZE Function

For our example, let's use the height of adult women in the U.S., measured in inches. Assume that the mean (μ) of the population is 64 inches (i.e., 5'4") and the standard deviation is 3 inches. Let's transform a height of 58 inches to a standardized value.

1. Type **X** in cell **A1**, type **Z** in cell **B1**, and type **58** in cell **A2**.

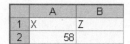

2. Activate cell **B2** in the output table where the Z for X = 58 inches will be placed.

3. Click **Insert** in the Menu Bar and select **Function**.

4. In the Insert Function dialog box, select the **Statistical** category and the **STANDARDIZE** function. Click **OK**.

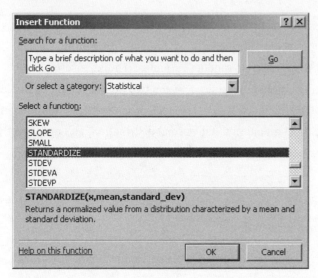

5. Complete the STANDARDIZE dialog box as shown below. Detailed instructions are given immediately following the dialog box.

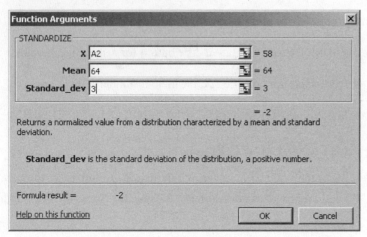

- **X**. Enter **A2**, the relative cell address for X equal to 58.

- **Mean**. Enter **64**, the population mean.

- **Standard_dev**. Enter **3,** the population standard deviation.

6. Click **OK** and a Z of -2 is returned.

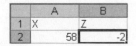

	A	B
1	X	Z
2	58	-2

Calculating Cumulative Probabilities Using the NORMSDIST Function

If we are working with variables that are distributed normally, one of the advantages of standardized scores is that we can easily determine the probability of obtaining a Z value that is less than, or greater than, a specified value. To illustrate the computation of cumulative probabilities, we will be using the NORMSDIST function. (If you are working with raw score values, you will use the NORMDIST function, which requires a couple more entries in the dialog box.)

The NORMSDIST works about the same as the normal distribution table found in the back of most statistics textbooks, only better. With textbook tables we are usually limited to Z values displayed with a maximum of two decimal places and must interpolate to obtain the probabilities associated with Z values with more than two decimals. To illustrate the beauty of Excel for the task of calculating cumulative probabilities, I have included in the example some Z values with four decimals, as well as a couple of extreme Z values not found in most textbook tables.

1. Prepare an output table by entering the column labels, **Z, P(Z<z), P(Z>z)**, and the specific Z values shown in column A below. Make columns B and C wide enough so that several decimal places can be displayed.

	A	B	C
1	Z	P(Z<z)	P(Z>z)
2	5.27		
3	2.6725		
4	1		
5	-1		
6	-2.6725		
7	-5.27		

2. **P(Z < 5.27)**. We will begin by calculating the probability of obtaining a Z value that is less than 5.27. Activate cell **B2** of the output table where the probability will be placed.

3. Click **Insert** in the Menu Bar and select **Function**.

4. In the Insert Function dialog box, select the **Statistical** category and the **NORMSDIST** function. Click **OK**.

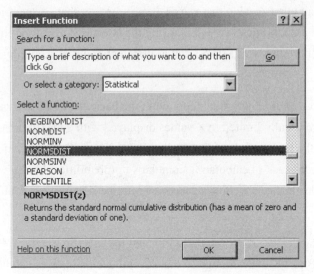

5. Click in the **Z** window and then click in cell **A2** of the worksheet to enter the worksheet address of 5.27 in the dialog box.

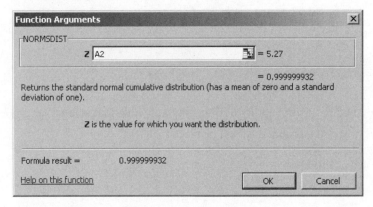

6. Click **OK** and .999999932 is returned. This is the probability of obtaining a Z value less than 5.27 in the standard normal cumulative distribution.

	A	B	C
1	Z	P(Z<z)	P(Z>z)
2	5.27	0.999999932	
3	2.6725		
4	1.96		
5	1.0548		
6	1		
7	0		
8	-1		
9	-1.0548		
10	-1.96		
11	-2.6725		
12	-5.27		

7. To obtain the cumulative probabilities for the remaining Z values, first activate cell **B2** where the cumulative probability associated with 5.27 was placed. Then use the copy command to copy the NORMSDIST function in cell B2 to cells B3 through B7. A completed output table is displayed below. As you can see, we can quickly obtain cumulative probabilities for familiar Z values such as 1, as well as for Z values with several decimal places (e.g., 2.6725).

	A	B	C
1	Z	P(Z<z)	P(Z>z)
2	5.27	0.999999932	
3	2.6725	0.996235581	
4	1	0.841344746	
5	-1	0.158655254	
6	-2.6725	0.003764419	
7	-5.27	6.82119E-08	

8. The column C probabilities are easy to obtain after completing column B. We will utilize the formula: 1 minus the value in column B. Activate cell **C2** where the first cumulative probability will be placed. Enter the formula **=1-B2** and press [**Enter**].

	A	B	C
1	Z	P(Z<z)	P(Z>z)
2	5.27	0.999999932	6.82119E-08
3	2.6725	0.996235581	
4	1	0.841344746	
5	-1	0.158655254	
6	-2.6725	0.003764419	
7	-5.27	6.82119E-08	

9. To obtain the remaining cumulative probabilities, copy the formula in cell C2 to cells C3 through C7. A completed table is shown below.

	A	B	C
1	Z	P(Z<z)	P(Z>z)
2	5.27	0.999999932	6.82119E-08
3	2.6725	0.996235581	0.003764419
4	1	0.841344746	0.158655254
5	-1	0.158655254	0.841344746
6	-2.6725	0.003764419	0.996235581
7	-5.27	6.82119E-08	0.999999932

Obtaining Z Values for Specified Probabilities Using the NORMSINV Function

Researchers sometimes want to know the Z value that defines a particular point in the distribution, such as the lower 10% or the upper 5%. To obtain these values, we can use the NORMSINV function. The procedure is quite straightforward, so I have selected only four values to illustrate the use of the NORMSINV function.

1. Begin by preparing an output table like the one shown below to display the probabilities and associated Z values.

	A	B
1	Probability	Z
2	Lower 1%	
3	Lower 5%	
4	Upper 5%	
5	Upper 1%	

2. **Lower 1%**. What Z value defines the lower 1% of the standard normal distribution? Activate cell **B2** where the Z value will be placed.

3. Click **Insert** in the Menu Bar and select **Function**.

4. In the Insert Function dialog box, select the **Statistical** category and the **NORMSINV** function. Click **OK**.

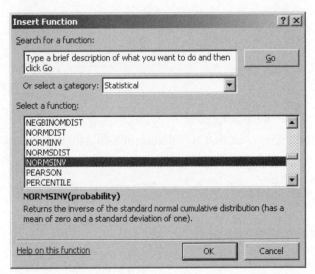

5. Probabilities are represented as proportions. For the lower 1%, enter **.01** in the Probability window.

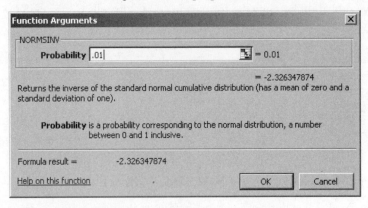

6. Click **OK** and a Z value of -2.3264 is returned.

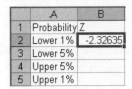

7. **Lower 5%**. What Z value defines the lower 5% of the standard normal distribution? Activate cell **B3** where the Z value will be placed.

8. Click **Insert** in the Menu Bar and select **Function**.

9. In the Insert Function dialog box, select the **Statistical** category and the **NORMSINV** function. Click **OK**.

10. For the lower 5%, enter **.05** in the Probability window.

11. Click **OK** and a value of -1.6449 is returned.

12. **Upper 5%**. What Z value defines the upper 5% of the standard normal distribution? Activate cell **B4** where the Z value will be placed.

13. Click **Insert** the Menu Bar and select **Function**.

14. In the Insert Function dialog box, select the **Statistical** category and the **NORMSINV** function. Click **OK**.

15. The point in the normal distribution that defines the upper 5% is the same point that defines the lower 95%. The NORMSINV is a cumulative probability function, so we need to enter .95 to obtain the correct Z value. Click in the Probability window and enter **.95**.

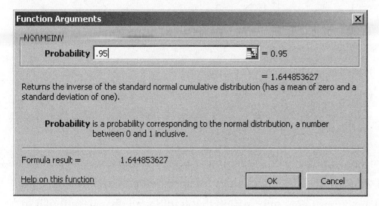

16. Click **OK** and a value of 1.6449 is returned.

17. **Upper 1%**. What Z value defines the upper 1% of the standard normal distribution? Activate cell **B5** where the Z value will be placed.

18. Click **Insert** in the Menu Bar and select **Function**.

19. In the Insert Function dialog box, select the **Statistical** category and the **NORMSINV** function. Click **OK**.

20.   The point in the distribution that defines the upper 1% also defines the lower 99%. Since NORMSINV
is a cumulative probability function, we need to enter .99 to obtain the correct Z value. Click in the
Probability window and enter **.99**.

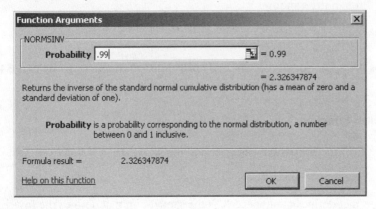

21.   Click **OK** and a value of 2.3264 is returned.

A completed output table is shown below.

	A	B
1	Probability	Z
2	Lower 1%	-2.32635
3	Lower 5%	-1.64485
4	Upper 5%	1.644854
5	Upper 1%	2.326348

t Distribution

In this chapter, I will limit my instructions regarding Excel's t distribution functions to the tasks of obtaining
probabilities associated with specified t values and obtaining t values for specified probabilities. In other
words, I will explain how to use Excel's functions to get information that traditionally has been found in
tables in statistics textbooks. I think you will agree with me that these functions are much better than the
tables.

Obtaining Probabilities for Specified t Values Using the TDIST Function

I selected six t values of varying magnitude to use as examples. These are displayed in the table shown
below. In addition to a column label for **t**, I also have columns for degrees of freedom (**df**), number of tails
(**Tails**), and **Probability**. When you use the TDIST function, you need to specify t, df, and number of tails.

	A	B	C	D
1	t	df	Tails	Probability
2	0.72	31	1	
3	0.72	31	2	
4	2.145	15	1	
5	2.145	15	2	
6	3.3125	63	2	
7	E.22143	124	2	

1.   **Output table**. Prepare an output table as shown above.

You might be wondering why the table contains only positive t values. If you request the probability of a negative t value, TDIST will return #NUM! The TDIST will only evaluate positive t values.

2. **Probability**. We will start with the first line in our output table, the 1-tailed probability of t = 0.72 in a distribution with df = 31. Activate cell **D2** where the probability will be placed.

3. Click **Insert** in the Menu Bar and select **Function**.

4. In the Insert Function dialog box, select the **Statistical** category and the **TDIST** function. Click **OK**.

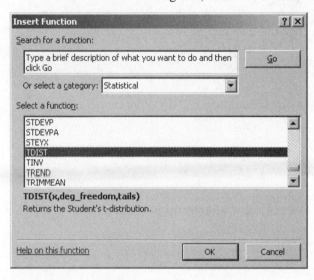

5. Complete the TDIST dialog box as shown below. A description of the entries is given immediately after the dialog box.

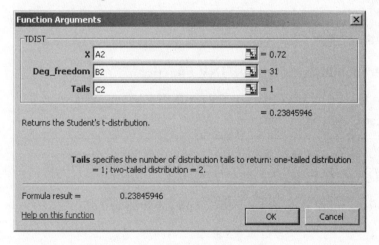

- **X**. X refers to the t value. Click in the **X** window and then click in **A2** where the first t value appears. If you prefer, you can manually enter **A2**.

> *Instead of cell addresses, you can enter numerical values in the TDIST dialog box. For example, you could enter .72, 31, and 1, respectively, in the X, Deg_freedom, and Tails windows of the TDIST dialog box. If you want the probability for only one t value, numerical entries might be faster than cell addresses. If you are working with two or more t values, however, the use of cell addresses enables you to obtain probabilities for the remaining t values by copying the contents of D2.*

- **Deg_freedom**. Click in the **Deg_freedom** window and then click in **B2** where the first df value appears. If you prefer, you can manually enter **B2**.

- **Tails**. Click in the **Tails** window and then click in **C2** where the first tails value appears.

6. Click **OK** and a probability of .2385 is returned.

	A	B	C	D
1	t	df	Tails	Probability
2	0.72	31	1	0.238459
3	0.72	31	2	
4	2.145	15	1	
5	2.145	15	2	
6	3.3125	63	2	
7	6.22143	124	2	

7. To obtain the probabilities for the remaining t values, copy the contents of D2 to cells D3 to D7.

A completed output table is shown below.

	A	B	C	D
1	t	df	Tails	Probability
2	0.72	31	1	0.238459
3	0.72	31	2	0.476919
4	2.145	15	1	0.024367
5	2.145	15	2	0.048734
6	3.3125	63	2	0.001535
7	6.22143	124	2	2.29E-13

Obtaining t Values for Specified Probabilities Using the TINV Function

Assume that a researcher wants to know the critical value of a 2-tailed t with df = 10 and alpha set at .05. To find the critical t value, the researcher can use the TINV function.

> *The TINV function returns only 2-tailed t values. If a 1-tailed t is needed, double your desired alpha (probability) and the correct value will be returned.*

I have set up an output table to display four critical t values for specified probabilities and df.

1. **Output table**. Prepare an output table like the one shown below.

	A	B	C
1	Probability	df	Critical t
2	0.05	10	

2. **Critical t**. We need to find the critical t value for probability (alpha) equal to .05 and df equal to 10. This is the critical t for two-tailed alpha equal to .05 or the critical t for one-tailed alpha equal to .025. Activate cell **C2** where the critical t value will be placed.

3. Click **Insert** in the Menu Bar and select **Function**.

4. In the Insert Function dialog box, select the **Statistical** category and the **TINV** function. Click **OK**.

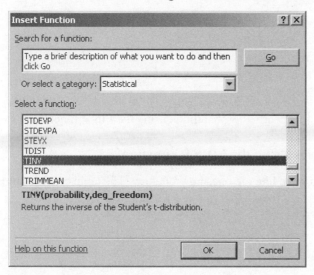

5. Complete the TINV dialog box as shown below. Information about the entries is given immediately after the dialog box.

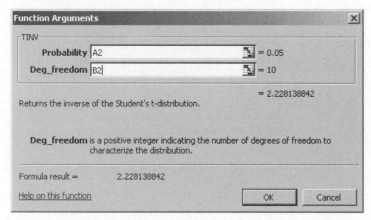

- Click in the **Probability** window and then click in **A2** of the worksheet where the first probability value is located. If you prefer, you can manually enter **A2**.

---

*Instead of cell addresses, you can enter numerical values in the TINV dialog box. For
example, you could enter .05 and 10, respectively, in the Probability and Deg_freedom
windows. If you want only one t value, numerical entries might be faster than cell addresses.
If you are working with two or more items, however, the use of cell addresses enables you to
obtain t values for the remaining probability and df combinations by copying the contents of
C2 .*

- Click in the **Deg_freedom** window and then click in cell **B2** where the first df value is located.

6. Click **OK** and a value of 2.2281 is returned.

	A	B	C
1	Probability	df	Critical t
2	0.05	10	2.228139

F Distribution

I will first explain how to find probabilities associated with specified F values using the FDIST function.
Then I will explain how to obtain critical F values for specified probabilities using the FINV function.

Obtaining Probabilities for Specified F Values Using the FDIST Function

1. **Output table**. Begin by preparing an output table like the one shown below. The table contains an F
    test value along with its df 1 (numerator degrees of freedom) and df 2 (denominator df).

	A	B	C	D
1	F	df 1	df 2	Probability
2	3.125	3	12	

2. **Probability**. Activate cell **D2** where the probability will be placed.

3. Click **Insert** in the Menu Bar and select **Function**.

4. In the Insert Function dialog box, select the **Statistical** category and the **FDIST** function. Click **OK**.

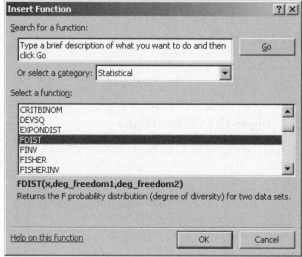

5. Complete the FDIST dialog box as shown below. Detailed information is given immediately following the dialog box.

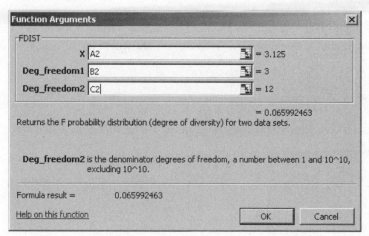

- **X** refers to the value of F for which you want to obtain the probability. Click in the **X** window and then click in **A2** where the F value is located.

> *Instead of cell addresses, you can enter numerical values in the FDIST dialog box. For example, you could enter 3.125, 3, and 12, respectively, in the X, Deg_freedom1, and Deg_freedom2 windows. If you want only one probability, numerical entries might be faster than cell addresses.   If you are working with two or more items, however, the use of cell addresses enables you to obtain probabilities for the remaining F value and df combinations by copying the contents of D2.*

- Deg_freedom1 refers to numerator df. Click in the **Deg_freedom1** window and then click in cell **B2** where df 1 is located.

- Deg_freedom2 refers to denominator df. Click in the **Deg_freedom2** window and then click in cell **C2** where df 2 is located.

6. Click **OK** and a probability of 0.0660 is returned.

	A	B	C	D
1	F	df 1	df 2	Probability
2	3.125	3	12	0.065992463

Obtaining F Values for Specified Probabilities Using the FINV Function

To obtain a critical F value, we will use the FINV function. We need to specify probability (alpha), numerator df (df 1), and denominator df (df 2).

1. **Output table**. We will find three critical F values. First set up an output table like the one shown below to display the probability, df 1, and df 2 for each critical F value.

	A	B	C	D
1	Probability	df 1	df 2	Critical F
2	0.05	2	80	
3	0.025	4	163	
4	0.002	6	58	

2. **Critical F**. Let's begin by finding the critical F in row 2. This F is distributed with 2 numerator df and 80 denominator df. The probability (alpha) is equal to .05. Activate cell **D2** where the critical F value will be placed.

3. Click **Insert** in the Menu Bar and select **Function**.

4. In the Insert Function dialog box, select the **Statistical** category and the **FINV** function. Click **OK**.

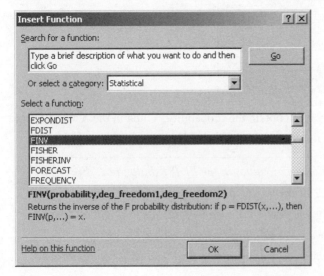

5. Complete the FINV dialog box as shown below. Detailed information regarding the entries is given immediately after the dialog box.

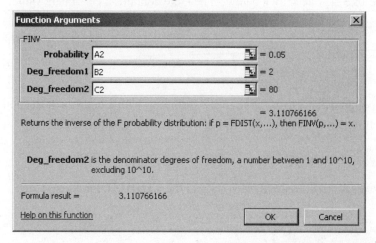

- Probability refers to the type I error probability or alpha. The probability for the first critical F in the output table is located in cell A2. So, click in the **Probability** window and then click in cell **A2**.

- Deg_freedom1 refers to numerator degrees of freedom. Click in the **Deg_freedom1** window and then click in cell **B2**.

- Deg_freedom2 refers to denominator degrees of freedom. Click in the **Deg_freedom2** window and then click in cell **C2**.

6. Click **OK** and a value of 3.1108 is returned.

	A	B	C	D
1	Probability	df 1	df 2	Critical F
2	0.05	2	80	3.110766
3	0.025	4	163	
4	0.002	6	58	

7. Copy cell D2 to cells D3 through D4. A completed output table is shown below.

	A	B	C	D
1	Probability	df 1	df 2	Critical F
2	0.05	2	80	3.110766
3	0.025	4	163	2.865226
4	0.002	6	58	4.009648

Chi-Square Distribution

I will first explain how to find probabilities associated with specified chi-square values using the CHIDIST function. Then I will explain how to obtain critical chi-square values for specified probabilities using the CHIINV function.

Obtaining Probabilities for Specified Chi-Square Values Using the CHIDIST Function

1. Begin by preparing an output table like the one shown below. The table contains a chi-square value, degrees of freedom (df), and probability.

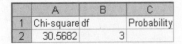

	A	B	C
1	Chi-square	df	Probability
2	30.5682	3	

2. We will find the probability of chi-square equal to 30.5682 in a distribution with 3 degrees of freedom. Click in cell **C2** where the probability will be placed.

3. Click **Insert** in the Menu Bar and select **Function**.

4. In the Insert Function dialog box, select the **Statistical** category and the **CHIDIST** function. Click **OK**.

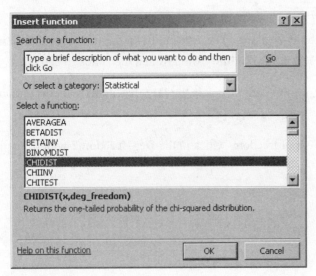

5. Complete the CHIDIST dialog box as shown below. Detailed information is given immediately following the dialog box.

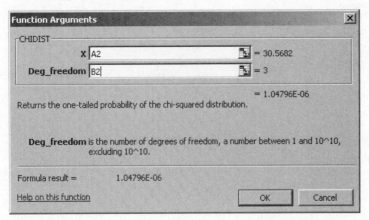

- X refers to the value of chi square for which you want to obtain the probability. Click in the **X** window and then click in cell **A2** of the worksheet where the chi square value is located.

- Deg_Freedom refers to the degrees freedom of the chi-square distribution you are working with. Click in the **Deg_Freedom** window and then click in cell **B2** of the worksheet where df is located.

6. Click **OK** and a probability of 1.04796E-06 is returned. Scientific notation is used in the output whenever the probability is very small.

	A	B	C
1	Chi-square	df	Probability
2	30.5682	3	1.04796E-06

Obtaining Chi-Square Values for Specified Probabilities Using the CHIINV Function

To obtain a critical chi-square value, we will use the CHIINV function. We need to specify probability (alpha) and degrees of freedom.

1. **Output table**. We will find three critical chi-square values. First set up an output table like the one shown below to display the probability and df for each critical chi-square value.

	A	B	C	D
1	Probability	df	Critical chi-square	
2	0.05	2		
3	0.01	2		
4	0.001	2		

2. Activate cell **C2** where you will place the critical value of chi-square in a distribution with 2 df and a probability of .05.

3. Click **Insert** in the Menu Bar and select **Function**.

4. In the Insert Function dialog box, select the **Statistical** category and the **CHIINV** function. Click **OK**.

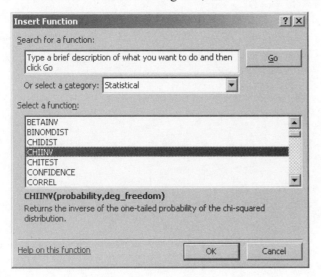

5. Complete the CHIINV dialog box as shown below. Detailed information regarding the entries is given immediately after the dialog box.

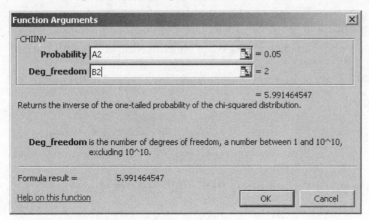

- Probability refers to the type I error probability (alpha). The probability of .05 was placed in cell A2. Click in the **Probability** window and then click cell **A2** of the worksheet.

- Deg_freedom refers to the degrees of freedom. The degrees of freedom were placed in cell B2. Click in the **Deg_freedom** window and then click in cell **B2** of the worksheet.

6. Click **OK** and a critical value of 5.9915 is returned.

	A	B	C	D
1	Probability	df	Critical chi-square	
2	0.05	2	5.991465	
3	0.01	2		
4	0.001	2		

7. Copy cell **C2** to cells C3 and C4 to obtain the critical chi-squares values for probabilities of .01 and .001, respectively. The completed output table is shown below.

	A	B	C	D
1	Probability	df	Critical chi-square	
2	0.05	2	5.991465	
3	0.01	2	9.21034	
4	0.001	2	13.81551	

Chapter 7

Testing Hypotheses About One Sample Means

If a researcher wants to compare the mean of one sample with a hypothesized population value, he or she should use the one-sample Z-test or the one-sample t-test. The selection will be based on the assumptions underlying the tests. For either test, the researcher will have one of three objectives in mind:

1. To test whether the sample mean is greater than the hypothesized population value,

2. To test whether the sample mean is less than the hypothesized population value, or

3. To test whether the sample mean is not equal to the hypothesized population value. That is to say, the sample mean could be either greater than or less than the hypothesized value.

Objectives 1 and 2 are associated with directional or one-tailed tests; objective 3 is associated with a nondirectional or two-tailed test. In this chapter, I first explain how to utilize Excel to carry out the one-sample Z-test, and then I explain the one-sample t-test. For both the Z-test and the t-test, I have included the one-tailed and two-tailed versions. I also describe how to set up confidence intervals for the nondirectional tests.

One-Sample Z-Test

The formula for the one-sample Z-test is

$$Z = \frac{\overline{X} - \mu}{\sigma_{\overline{X}}}$$

where $\overline{X}$ is the sample mean, μ is the hypothesized value of the population mean, and $\sigma_{\overline{X}}$ is the standard error of the mean.

The formula for the standard error of the mean is given by

$$\sigma_{\overline{X}} = \frac{\sigma}{\sqrt{n}}$$

where σ is the standard deviation of the population, and n is the number of subjects in the sample.

Assumptions Underlying the Z-Test

The statistical assumptions underlying the one-sample Z-test include:

1. Observations are independent of one another.

2. The observations are randomly sampled from the population.

3. Observations are normally distributed in the population.

4. The population variance, σ^2, is known.

Sample Research Problem

The increase in the number of children who are schooled at home has stimulated numerous research questions regarding the characteristics of these children. One such question is related to their level of intelligence. Do home-schooled children possess the same average level of intelligence as other children?   It might be assumed that parents who decide to educate their children at home are fairly bright, and that their children, in turn, are also fairly bright. A researcher who was interested in this question randomly selected 12 seven-year-olds who were being home-schooled and administered to each child the Wechsler Intelligence Scale for Children (WISC). The published norms for the WISC indicate that the scores for the population are normally distributed with a mean equal to 100 and a standard deviation equal to 15.   Is the mean score on the WISC for seven-year-old home-schooled children significantly different from 100?

Steps to Follow to Analyze the Sample Research Problem

1. Enter the data in an Excel worksheet as shown below.

	A
1	WISC
2	97
3	112
4	119
5	84
6	135
7	95
8	127
9	103
10	95
11	107
12	101
13	98

2. Prepare an output table in the same worksheet as the data to display a summary of the data analysis. In columns C and D, enter the labels and numbers as shown below. Be sure to use exactly the same cell locations as I have, because I frequently refer to cell locations in the instructions. You may need to make column C wider to accommodate the long labels.

	A	B	C	D
1	WISC		ONE-SAMPLE Z-TEST	
2	97		Sample mean	
3	112		Hypothesized population mean	100
4	119		Population standard deviation	15
5	84		Count	
6	135		Standard error of the mean	
7	95		Z	
8	127		Alpha	0.05
9	103		Probability one-tailed	
10	95		Z critical one-tailed	
11	107		Probability two-tailed	
12	101		Z critical two-tailed	
13	98			

3. **Sample mean.** Use the AVERAGE function to obtain the sample mean. Activate cell **D2**. Click **Insert** in the Menu Bar and select **Function**.

4. In the Insert Function dialog box, select the **Statistical** category. Select **AVERAGE** in the function list. Click **OK**.

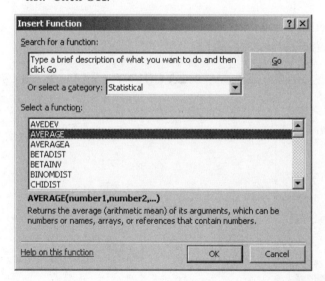

5.   You now need to tell Excel where the data are located. Click in the **Number 1** window and remove the information, if any, that appears there. Then click in the top cell of the WISC data (cell **A2**), and drag to the end of the data (cell **A13**). If you prefer, you can manually enter **A2:A13** in the Number 1 window.

The AVERAGE function will also give you an accurate result if you include cell A1 (WISC) in the Number 1 range.

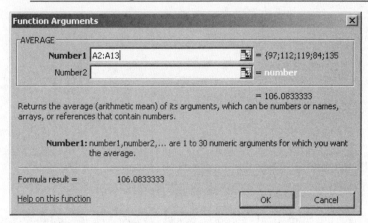

6.   Click **OK** and 106.0833 will appear in cell D2 of your worksheet.

7.   **Count**. Next you will use the COUNT function to count the number of observations in the sample. Activate cell **D5**. Click **Insert** in the Menu Bar and select **Function**.

8.   In the Insert Function dialog box, select the **Statistical** category. Select **COUNT** in the function list. Click **OK**.

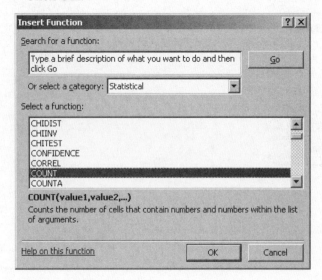

9. Delete information, if any, that appears in the Value 1 window. Then click in the **Value 1** window, click in the top cell of WISC data (**A2**), and drag to the end of the data (**A13**). If you prefer, you can manually enter **A2:A13** in the Value 1 window.

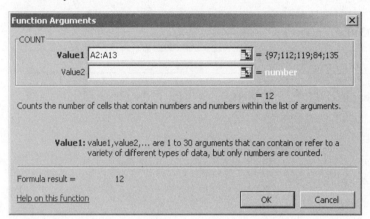

10. Click **OK** and 12 will appear in cell D5 of the worksheet.

	A	B	C	D
1	WISC		ONE-SAMPLE Z-TEST	
2	97		Sample mean	106.0833
3	112		Hypothesized population mean	100
4	119		Population standard deviation	15
5	84		Count	12
6	135		Standard error of the mean	
7	95		Z	
8	127		Alpha	0.05
9	103		Probability one-tailed	
10	95		Z critical one-tailed	
11	107		Probability two-tailed	
12	101		Z critical two-tailed	
13	98			

11. **Standard error of the mean**. The formula for the standard error of the mean is

$$\sigma_{\overline{X}} = \frac{\sigma}{\sqrt{n}}$$

You will use this formula to calculate the standard error. Activate cell **D6** where you will place the value of the standard error.

---

*The formula you will be entering in cell D6 is =**D4/SQRT(D5)**. The equal sign tells Excel that the information that follows will be a formula. D4 is the cell location of σ, and D5 is the cell location of n (Count). The diagonal indicates division. SQRT is the abbreviation for the square root function. Excel will compute the square root of the number in parentheses immediately following SQRT.*

---

12. Enter =**D4/SQRT(D5)** in cell D6. Press [**Enter**] and a value of 4.3301 will be returned.

13.   **Z**. You will now key in another formula to calculate Z. Recall that the formula for Z is

$$Z = \frac{\overline{X} - \mu}{\sigma_{\overline{X}}}$$

Activate cell **D7** and enter **=(D2-D3)/D6**.

The period after D6 is not part of the formula.

14.   Press [**Enter**] and 1.4049 will appear in cell D7 of the worksheet. This is the obtained value of Z.

	A	B	C	D
1	WISC		ONE-SAMPLE Z-TEST	
2	97		Sample mean	106.0833
3	112		Hypothesized population mean	100
4	119		Population standard deviation	15
5	84		Count	12
6	135		Standard error of the mean	4.330127
7	95		Z	1.404886
8	127		Alpha	0.05
9	103		Probability one-tailed	
10	95		Z critical one-tailed	
11	107		Probability two-tailed	
12	101		Z critical two-tailed	
13	98			

15.   **Probability one-tailed.** Next, you will find the one-tailed probability associated with the obtained value of Z. Activate cell **D9** where the one-tailed probability will be displayed. You will be using a formula to obtain the one-tailed probability. When completed, it will look like:

=1-NORMSDIST(ABS(D7))

NORMSDIST is the standard normal cumulative distribution function. It returns the cumulative probability associated with a given Z value. ABS is the absolute value function. Because it is possible for Z to take on negative as well as positive values, the task of evaluating the statistical significance of Z will be simplified if you are working with an absolute value. Finally, D7 is the cell location of Z.

16.   Enter **=1-** and then click **Insert** in the Menu Bar and select **Function**.

17. In the Insert Function dialog box, select the **Statistical** category. Select **NORMSDIST** in the function list. Click **OK**.

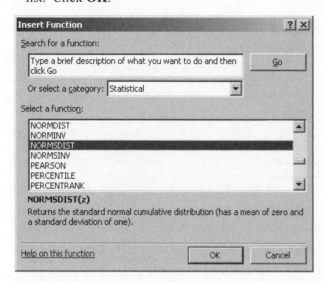

18. Enter **ABS(D7)** in the Z window and click **OK**. A value of .0800, the one-tailed probability of obtained Z, will appear in cell D9 of the worksheet.

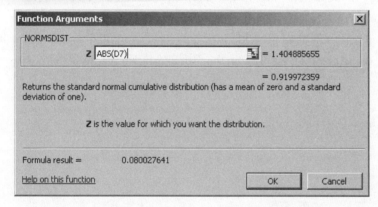

19. **Z critical one-tailed**. Activate cell **D10**. The formula that you will be entering is:

=ABS(NORMSINV(.05))

This formula contains two functions. ABS is the absolute value function. NORMSINV is the inverse of the standard normal cumulative distribution function. NORMSINV provides the Z value associated with a specified probability. You want the Z value associated with a probability of .05, the selected value of alpha.

20. Key in the formula **=ABS(NORMSINV(.05))** and press [**Enter**]. A critical Z of 1.6449 will be displayed in the output table.

	A	B	C	D
1	WISC		ONE-SAMPLE Z-TEST	
2	97		Sample mean	106.0833
3	112		Hypothesized population mean	100
4	119		Population standard deviation	15
5	84		Count	12
6	135		Standard error of the mean	4.330127
7	95		Z	1.404886
8	127		Alpha	0.05
9	103		Probability one-tailed	0.080028
10	95		Z critical one-tailed	1.644854
11	107		Probability two-tailed	
12	101		Z critical two-tailed	
13	98			

21. **Probability two-tailed.** Because the two-tailed probability can be obtained by doubling the number shown for the one-tailed probability, we will enter a simple formula to compute the two-tailed probability using the number displayed in cell D9. Activate cell **D11** where you will place the two-tailed probability and key in **=2*D9**. Press [**Enter**] and .1601 should appear in your worksheet.

22. **Z critical two-tailed.** The last entry in the table is the two-tailed critical value of Z. To obtain this value, you will enter a formula like the one you used for the one-tailed critical value, with one exception. Alpha is equally divided between the upper and lower tails for a two-tailed test, so you will use .025 in the formula instead of .05. Activate **D12**. Key in the formula shown below.

    =ABS(NORMSINV(.025))

23. Press [**Enter**] and 1.9600 will be displayed in the table.

Interpreting the Output

	A	B	C	D
1	WISC		ONE-SAMPLE Z-TEST	
2	97		Sample mean	106.0833
3	112		Hypothesized population mean	100
4	119		Population standard deviation	15
5	84		Count	12
6	135		Standard error of the mean	4.330127
7	95		Z	1.404886
8	127		Alpha	0.05
9	103		Probability one-tailed	0.080028
10	95		Z critical one-tailed	1.644854
11	107		Probability two-tailed	0.160055
12	101		Z critical two-tailed	1.959964
13	98			

- **Sample mean.** The mean WISC score for the 12 children in the sample, $\overline{X} = 106.0833$.

- **Hypothesized population mean.** The value 100 indicates that the hypothesis being tested in this analysis is that the population mean of children who are home-schooled is equal to 100.

- **Population standard deviation.** The population standard deviation (or the population variance) must be known in order for the Z-test to be applied appropriately to a set of data. For this analysis, the population standard deviation is equal to 15.

- **Count.** The number of observations in the sample (n). For this problem, n is equal to 12.

- **Standard error of the mean.** The standard error is the standard deviation of the sampling distribution of the mean, and, as such, it provides an approximation of the average amount by which the sample means deviate from the population mean. For this example, you can say that the distribution of sample means for samples of size 12 drawn from the population of WISC scores for home-schooled children has a standard deviation of 4.3301 points.

- **Z.** Value of Z obtained by applying the formula presented at the beginning of this section. For this research problem, obtained Z equals 1.4049.

- **Alpha.** The significance level for the statistical test. This value is selected by the researcher(s), and for this analysis, we selected .05.

- **Probability one-tailed.** One-tailed probability of the obtained Z, also referred to as the one tailed P value. For this problem, if the null hypothesis is true that the population mean WISC score for home-schooled children is 100, then the chance probability of obtaining a Z ≤ −1.4049 is .0800. Or, in the upper tail of the distribution, the chance probability of obtaining a Z ≥ +1.4049 is .0800. Because the probability of obtained Z is greater than alpha, you would conclude that the result is not statistically significant with a one-tailed test.

- **Z critical one-tailed.** Absolute value of the one-tailed critical value of Z for the selected value of alpha. With alpha equal to .05, and an alternative hypothesis that says that the population mean WISC score for home-schooled children is less than 100, the one-tailed critical value of Z is −1.6445. Similarly, for an alternative hypothesis that says that the population mean WISC score for home-schooled children is greater than 100, the one-tailed critical value of Z is +1.6445. Assuming that your alternative hypothesis stated that the population mean was greater than 100, you would have to conclude that the one-tailed test is not statistically significant because the obtained Z of 1.4049 is not greater than the critical Z.

- **Probability two-tailed.** Two-tailed chance probability (or two-tailed P-value) associated with the obtained Z. More specifically, if the null hypothesis is true that the population mean WISC score for home-schooled children is 100, then the chance probability of obtaining a Z ≤ −1.4049 plus the chance probability of obtaining a Z ≥ + 1.4049 is .1601. Note that this value is two times the one-tailed probability of .0800. Because .1601 is greater than alpha (.05), you would conclude that the two-tailed test is not statistically significant.

- **Z critical two-tailed.** Absolute value of the two-tailed critical value of Z for the selected value of alpha. For alpha equal to .05, the two-tailed critical value of Z is 1.9600. This result is not statistically significant, because obtained Z (1.4049) is less than critical Z.

Confidence Interval for the One-Sample Z-Test

To calculate the lower and upper limits of the confidence interval, we use the following general expression:

$$\overline{X} \pm (Z_{Crit})(\sigma_{\overline{X}})$$

where $\overline{X}$ represents the sample mean, Z_{Crit} represents the two-tailed Z critical value, and $\sigma_{\overline{X}}$ represents the standard error of the mean. We will be using information that is provided in the ONE-SAMPLE Z-TEST output table. For easy reference, let's place the confidence interval output just below that table.

1. **Labels**. Beginning in cell C14, key in the following labels: **CONFIDENCE INTERVAL**, **Lower limit**, and **Upper limit**. Be sure to use the same cell placement as shown below, because I refer to these cell locations in the instructions.

11	107		Probability two-tailed	0.160055
12	101		Z critical two-tailed	1.959964
13	98			
14			CONFIDENCE INTERVAL	
15			Lower Limit	
16			Upper Limit	

2. **Lower limit**. Activate cell **D15**. Enter **=D2-(D12*D6)**.

---

*In this formula, D2 is the cell location of the sample mean, D12 is the cell location of the two-tailed Z critical value, * indicates multiplication, and D6 is the cell location of the standard error of the mean.*

---

3. Press [**Enter**]. You should now see 97.5964 in cell D15 of the output table.

14			CONFIDENCE INTERVAL	
15			Lower Limit	97.59644
16			Upper Limit	

4. **Upper limit**. The formula for the upper limit is nearly identical to the formula for the lower limit, except that you will be adding instead of subtracting. Activate cell **D16** and enter **=D2+(D12*D6)**.

5. Press [**Enter**] and 114.5702 should appear in cell D16 of the output table.

14			CONFIDENCE INTERVAL	
15			Lower Limit	97.59644
16			Upper Limit	114.5702

Interpreting the Confidence Interval

The confidence interval for a population mean specifies the range of values that, with a known degree of confidence, includes the unknown population mean. The computing formula for the confidence interval contains two elements: an estimate of the unknown population mean, and estimation error. If the test result is statistically significant, then the sample mean provides a good estimate of the unknown population mean. Estimation error is represented by the product, $(Z_{Crit})(\sigma_{\overline{X}})$. When alpha is equal to .05, a 95% confidence interval will be constructed. The 95% confidence interval for our research problem would take the form:

$97.5964 \le \mu \le 114.5702$. Based on this interval, we might claim that we are 95% confident that the true population mean lies between 97.5964 and 114.5702. More specifically, if we repeated the data analysis many times, sampling all possible samples of n equals 12 children from the population of home-schooled children, 95% of the confidence intervals would include the true population mean. The remaining 5% would not. Note that the confidence interval is consistent with an obtained Z-test value that was not statistically significant, because the hypothesized value of the population mean (100) is included in the interval.

One-Sample t-Test

The formula for the one-sample t-test is given by

$$t = \frac{\overline{X} - \mu}{S_{\overline{X}}}$$

where $\overline{X}$ is the sample mean, μ is the hypothesized value of the population mean, and $S_{\overline{X}}$ is the estimated standard error of the mean.

The formula for the estimated standard error of the mean is given by

$$S_{\overline{X}} = \frac{S}{\sqrt{n}}$$

where S is the standard deviation of the sample and n is the number of observations included in the sample.

Assumptions Underlying the t-Test

The statistical assumptions underlying the one-sample t-test are as follows:

1. Observations are independent of one another.

2. The observations are randomly sampled from the population.

3. Observations are normally distributed in the population.

4. The population variance, σ^2, is not known.

Sample Research Problem

It was recently reported that adults in the United States spend an average of 28 hours each week watching TV. A media researcher wanted to find out if adults in other countries also watched TV an average of 28 hours each week. The researcher asked 15 randomly selected Canadian adults to maintain careful records of their TV viewing time for one week.

Steps to Follow to Analyze the Sample Research Problem

1. Enter the data in column A of an Excel worksheet as shown at the top of the next page. Enter the labels in column C. Enter the numerical information in column D. Be sure to use exactly the same cell

locations, because cell addresses appear frequently in the instructions. You may want to make column C wider so that the longest label will be displayed in its entirety.

	A	B	C	D
1	HOURS		ONE-SAMPLE T-TEST	
2	15		Sample mean	
3	7		Hypothesized population mean	28
4	32		Sample standard deviation	
5	26		Count	
6	31		Standard error of the mean	
7	5		t	
8	0		Alpha	0.01
9	14		df	
10	11		Probability one-tailed	
11	3		t critical one-tailed	
12	38		Probability two-tailed	
13	10		t critical two-tailed	
14	9			
15	6			
16	11			

2. **Sample mean**. You will obtain the value of the sample mean by using the AVERAGE function. Activate cell **D2**. Click **Insert** in the Main Menu bar and selection **Function**.

3. In the Insert Function dialog box, select the **Statistical** category. Select **AVERAGE** in the function list. Click **OK**.

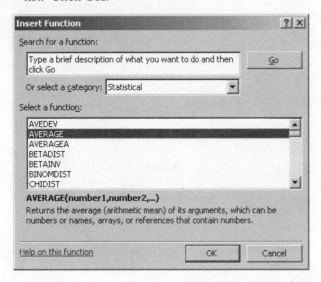

4. Delete the information, if any, that appears in the **Number 1** window. Then click in the top cell of the HOURS data (**A2**) and drag to the end of the data (**A16**). If you prefer, you can manually enter **A2:A16** in the Number 1 window.

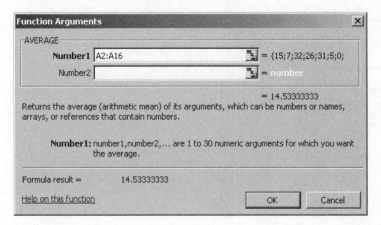

5. Click **OK** and 14.5333 should appear in cell D2 of the worksheet.

6. **Sample standard deviation**. Use the STDEV function to obtain the standard deviation of the sample data. Activate cell **D4**. Click **Insert** in the Menu Bar and select **Function**.

The STDEV function uses the unbiased estimate formula, which is given by

$$S = \sqrt{\frac{\sum (X - \overline{X})^2}{n - 1}}$$

---

7. In the Insert Function dialog box, select the **Statistical** category. Select **STDEV** in the function list. Click **OK**.

8. Delete information, if any, that appears in the **Number 1** window. Then click in the top cell of the HOURS data (**A2**) and drag to the end of the data (**A16**). If you prefer, you can manually enter **A2:A16** in the Number 1 window.

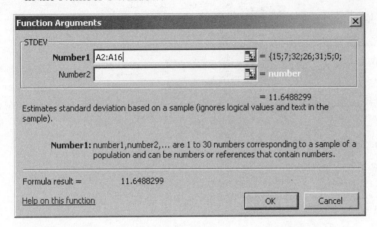

9. Click **OK** and 11.6488 will appear in cell D4 of the output table.

	A	B	C	D
1	HOURS		ONE-SAMPLE T-TEST	
2	15		Sample mean	14.53333
3	7		Hypothesized population mean	28
4	32		Sample standard deviation	11.64883
5	26		Count	
6	31		Standard error of the mean	
7	5		t	
8	0		Alpha	0.01
9	14		df	
10	11		Probability one-tailed	
11	3		t critical one-tailed	
12	38		Probability two-tailed	
13	10		t critical two-tailed	
14	9			
15	6			
16	11			

10. **Count**. Use the COUNT function to obtain a count of the observations in the sample. Activate cell **D5**. Click **Insert** in the Menu Bar and select **Function**.

11. In the Insert Function dialog box, select the **Statistical** category. Select **COUNT** in the function list. Click **OK**.

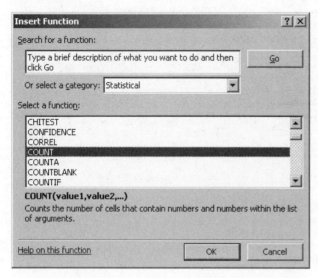

12. Delete information, if any, that appears in the **Value 1** window. Then click in the top cell of the HOURS data (**A2**) and drag to the end of the data (**A16**). If you prefer, you can manually enter **A2:A16** in the Value 1 window.

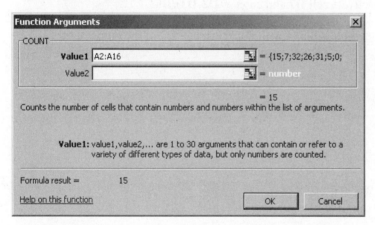

13. Click **OK** and 15 will appear in cell D5 of the worksheet.

14. **Standard error of the mean**. The formula for the standard error of the mean for the t-test is given by

$$S_{\overline{X}} = \frac{S}{\sqrt{n}}$$

Activate cell **D6** in the output table. Key in the formula **=D4/SQRT(D5)**.

The equal sign tells Excel that the information that follows will be a formula. D4 is the cell address of S, the diagonal indicates division, SQRT is the abbreviation for the square root function, and D5 is the cell address of n.

---

15.   Press **[Enter]** and 3.0077, the value of the standard error of the mean, will be displayed in cell D6 of the output table.

	A	B	C	D
1	HOURS		ONE-SAMPLE T-TEST	
2	15		Sample mean	14.53333
3	7		Hypothesized population mean	28
4	32		Sample standard deviation	11.64883
5	26		Count	15
6	31		Standard error of the mean	3.007715
7	5		t	
8	0		Alpha	0.01
9	14		df	
10	11		Probability one-tailed	
11	3		t critical one-tailed	
12	38		Probability two-tailed	
13	10		t critical two-tailed	
14	9			
15	6			
16	11			

16.   **t.** The formula for the one-sample t-test is given by

$$t = \frac{\overline{X} - \mu}{S_{\overline{X}}}$$

To apply this formula to the data, activate cell **D7** and enter **=(D2-D3)/D6**.

17.   Press **[Enter]** and the obtained value of t, −4.4774, will appear in cell D7 of the output table.

18. **df**. The formula for degrees of freedom (df) for the one-sample t-test is n – 1.   With n equal to 15, df = 14.   Enter **14** in cell **D9** of the output table.

	A	B	C	D
1	HOURS		ONE-SAMPLE T-TEST	
2	15		Sample mean	14.53333
3	7		Hypothesized population mean	28
4	32		Sample standard deviation	11.64883
5	26		Count	15
6	31		Standard error of the mean	3.007715
7	5		t	-4.47737
8	0		Alpha	0.01
9	14		df	14
10	11		Probability one-tailed	
11	3		t critical one-tailed	
12	38		Probability two-tailed	
13	10		t critical two-tailed	
14	9			
15	6			
16	11			

19. **Probability one-tailed**. You will use the **TDIST** function to find the one-tailed probability of obtained t (–4.4774). Activate cell **D10** in the output table. Click on **Insert** in the Menu bar and select **Function**.

20. In the Insert Function dialog box, select the **Statistical** category and the **TDIST** function. Click **OK**.

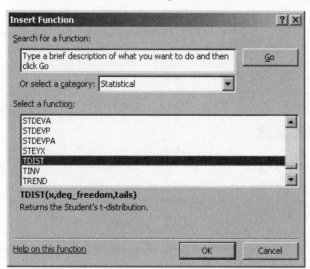

21.   Complete the TDIST dialog box as shown below. An explanation of the entries is given immediately after the dialog box.

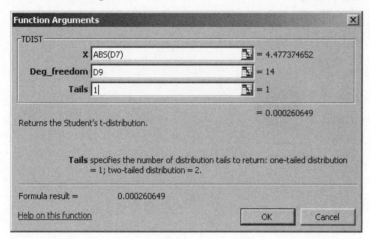

- Be certain that the flashing I-beam is positioned in the **X** window and key in **ABS(D7)**.

   ABS is the absolute value function, and D7 is the address of the obtained t value. Note that you need the absolute value function here because the TDIST function is not designed to evaluate negative values.

- Click in the **Deg_freedom** window.   Then click in cell **D9** to place the value of 14 in the window. If you prefer, you can manually enter **D9** or **14** in the window.

- Click in the **Tails** window. Enter **1** to obtain the one-tailed probability.

22.   Click **OK**. The one-tailed probability of .0003 should now be displayed in cell D10 of the output table.

	A	B	C	D
1	HOURS		ONE-SAMPLE T-TEST	
2	15		Sample mean	14.53333
3	7		Hypothesized population mean	28
4	32		Sample standard deviation	11.64883
5	26		Count	15
6	31		Standard error of the mean	3.007715
7	5		t	-4.47737
8	0		Alpha	0.01
9	14		df	14
10	11		Probability one-tailed	0.000261
11	3		t critical one-tailed	
12	38		Probability two-tailed	
13	10		t critical two-tailed	
14	9			
15	6			
16	11			

23.   **t critical one-tailed**. You will find the one-tailed critical value of t by using the **TINV** function. Activate cell **D11** in the output table. Click **Insert** in the Menu Bar and select **Function**.

TINV is designed to return the two-tailed critical value for specified values of alpha and df. In order to use this function for a one-tailed test, you need to double the value selected for alpha (i.e., 2 X .01 = .02) so that you will have .01 in each tail.

24.   In the Insert Function dialog box, select the **Statistical** category and the **TINV** function. Click **OK**.

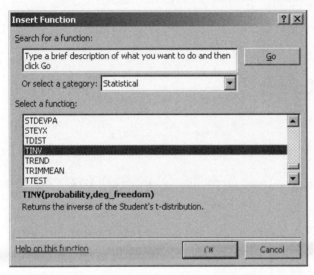

25.   Click in the **Probability** window and enter **.02**.

26.   Click in the **Deg_freedom** window and then click in cell **D9** in the output table to place the cell address of the df in the window. If you prefer, you can manually enter **D9** or **14** in the window.

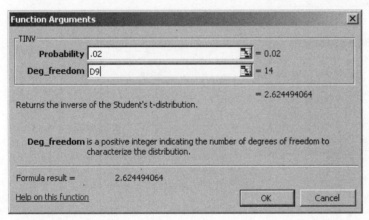

27.   Click **OK**. A critical one-tailed t of 2.6245 will now be displayed in cell D11 of the output table.

	A	B	C	D
1	HOURS		ONE-SAMPLE T-TEST	
2	15		Sample mean	14.53333
3	7		Hypothesized population mean	28
4	32		Sample standard deviation	11.64883
5	26		Count	15
6	31		Standard error of the mean	3.007715
7	5		t	-4.47737
8	0		Alpha	0.01
9	14		df	14
10	11		Probability one-tailed	0.000261
11	3		t critical one-tailed	2.624494
12	38		Probability two-tailed	
13	10		t critical two-tailed	
14	9			
15	6			
16	11			

28.   You will utilize the TDIST function to find the two-tailed probability. Activate cell **D12**. Click **Insert** in the Menu Bar and select **Function**.

29.   In the Insert Function dialog box, select the **Statistical** category and the **TDIST** function. Click **OK**.

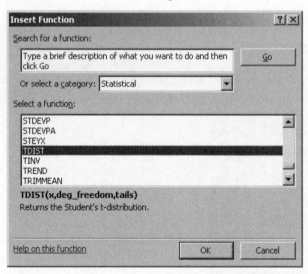

30. Complete the TDIST dialog box as shown below. An explanation of the entries is given immediately following the dialog box.

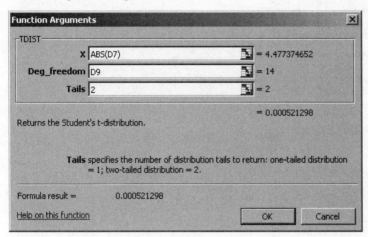

- Make sure that the flashing I-beam I located in the **X** window. Then enter **ABS(D7)**.

   ABS is the absolute value function, and D7 is the address of the obtained t value. Note that you need the absolute value function here because the TDIST function is not designed to evaluate negative numbers.

- Click in the **Deg_freedom** window and then click in cell **D9**. If you prefer, you can manually enter **D9** or **14** in the window.

- Click in the **Tails** window and enter **2** to obtain a two-tailed probability.

31. Click **OK**. A probability of .0005 is returned and placed in cell D12 of the output table.

	A	B	C	D
1	HOURS		ONE-SAMPLE T-TEST	
2	15		Sample mean	14.53333
3	7		Hypothesized population mean	28
4	32		Sample standard deviation	11.64883
5	26		Count	15
6	31		Standard error of the mean	3.007715
7	5		t	-4.47737
8	0		Alpha	0.01
9	14		df	14
10	11		Probability one-tailed	0.000261
11	3		t critical one-tailed	2.624494
12	38		Probability two-tailed	0.000521
13	10		t critical two-tailed	
14	9			
15	6			
16	11			

32. **t critical two-tailed**. You will use the TINV function to obtain the two-tailed critical value of t. Activate cell **D13**. Click **Insert** in the Menu Bar and selection **Function**.

33. In the Insert Function dialog box, select the **Statistical** category and the **TINV** function. Click **OK**.

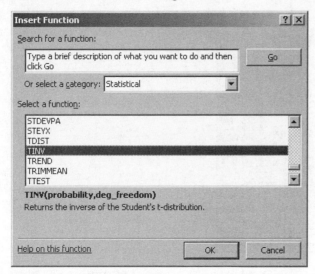

34. Click in the **Probability** window and then click in cell **D8**, the cell containing alpha. Click in the **Deg_freedom** window and then click in cell **D9**, the cell containing df.

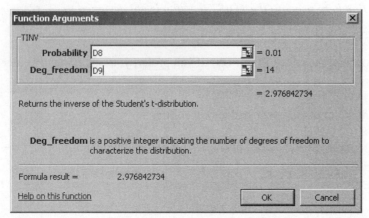

35. Click **OK** and the two-tailed critical t of 2.9768 will appear in the output table.

Interpreting the Output

	A	B	C	D
1	HOURS		ONE-SAMPLE T-TEST	
2	15		Sample mean	14.53333
3	7		Hypothesized population mean	28
4	32		Sample standard deviation	11.64883
5	26		Count	15
6	31		Standard error of the mean	3.007715
7	5		t	-4.47737
8	0		Alpha	0.01
9	14		df	14
10	11		Probability one-tailed	0.000261
11	3		t critical one-tailed	2.624494
12	38		Probability two-tailed	0.000521
13	10		t critical two-tailed	2.976843
14	9			
15	6			
16	11			

- **Sample mean**. The mean hours of weekly TV viewing for the persons included in the sample, $\overline{X} = 14.5333$.

- **Hypothesized population mean**. The value 28 indicates that the hypothesis under test in this analysis is that the population mean viewing time for adults residing in Canada is 28 hours.

- **Sample standard deviation**. The standard deviation of the sample, S = 11.6488. This value was computed using the formula for the unbiased estimate of the population standard deviation.

- **Count**. The number of observations in the sample, n = 15.

- **Standard error of the mean**. The standard error of the mean is the standard deviation of the sampling distribution of the mean. It provides an approximate average amount by which sample means differ from the population mean. For our research problem, we can say that the distribution of means for samples of n = 15 taken from the population of weekly TV viewing times of Canadian adults has a standard deviation of 3.0077.

- **t**. The value of t obtained by applying the formula shown at the beginning of this section. For this problem, obtained t is equal to -4.4774.

- **Alpha**. The significance level for the statistical test. This value is selected by the researcher. Because we wanted to perform a relatively conservative test, we set alpha at .01 for this analysis.

- **df**. The degrees of freedom for the statistical test. For the one-sample t-test, the formula for df is n − 1. With n equal to 15, df = 14.

- **Probability one-tailed**. One-tailed chance probability of obtained t. More specifically, if the null hypothesis is true that the population mean TV viewing time is 28 hours per week, then the chance probability of obtaining a t ≤ − 4.4774 is .0003. Similarly, for a value in the upper tail of the distribution, the chance probability of obtaining a t ≥ +4.4774 is .0003. Because the one-tailed probability of obtained t is less than alpha (.01), the one-tailed test would be considered statistically significant—assuming that the direction of the difference is consistent with the researcher's alternative hypothesis.

- **t critical one-tailed**. Absolute value of the one-tailed critical value of t for the selected value of alpha. With alpha equal to .01, and an alternative hypothesis that says that the population mean number of TV viewing hours for Canadian adults is less than 28, the one-tailed critical value of t is –2.6245. Similarly, for an alternative hypothesis that says that the population mean for Canadian adults is greater than 28, the one-tailed critical value of t is +2.6245. If the alternative hypothesis stated that the mean for the Canadians was less than 28 hours, the one-tailed test would be declared statistically significant, because the absolute value of obtained t (4.4774) is greater than the absolute value of the one-tailed critical t.

- **Probability two-tailed**. Two-tailed chance probability associated with the obtained t. More specifically, if the null hypothesis is true that the population mean TV viewing time for Canadian adults is 28 hours, then the chance probability of obtaining a t $\leq$ – 4.4774 plus the chance probability of obtaining a t $\geq$ +4.4774 is .0005. Note that this value is double that of the one-tailed probability. Because the two-tailed probability is less than alpha (.01), we would conclude that the two-tailed test is statistically significant.

- **t critical two-tailed**. Absolute value of the two-tailed critical value of t for the selected value of alpha. For alpha equal to .01, the two-tailed critical value of t is 2.9768. Since the absolute value of obtained t (4.4774) is greater than the absolute value of critical t, the two-tailed test result would be declared statistically significant.

Confidence Interval for the One-Sample t-Test

To compute the upper and lower limits of the confidence interval for the one-sample t-test, we use the following general expression:

$$\overline{X} \pm (t_{Crit})(S_{\overline{X}})$$

where $\overline{X}$ represents the sample mean, t_{Crit} represents the two-tailed t critical value, and $S_{\overline{X}}$ represents the standard error of the mean. To calculate the limits of the confidence interval, we will use the information provided in the ONE-SAMPLE T-TEST output table. We begin by setting up an output table to display a summary of our work.

1. **Labels**. For convenient reference, we'll place the CONFIDENCE INTERVAL output below the ONE-SAMPLE T-TEST output table. As shown below, begin in cell C15 and enter the labels:
   CONFIDENCE INTERVAL, Lower limit, Upper limit.

12	38	Probability two-tailed	0.000521
13	10	t critical two-tailed	2.976843
14	9		
15	6	CONFIDENCE INTERVAL	
16	11	Lower Limit	
17		Upper Limit	

2. **Lower limit**. We will use the following formula to calculate the lower limit of the confidence interval: **=D2-(D13*D6)**. Activate cell **D16** and key in **=D2-(D13*D6)**. Press [**Enter**] and the calculated value of the lower limit, 5.5798, will appear in the output table.

---

*In this expression, D2 is the cell location of the sample mean, D13 is the cell location of the two-tailed t critical value, * indicates multiplication, and D6 is the cell location of the standard error of the mean.*

15	6	CONFIDENCE INTERVAL	
16	11	Lower Limit	5.579839
17		Upper Limit	

3. **Upper limit**. The formula you will enter for the upper limit of the confidence interval is **=D2+(D13*D6)**. Activate cell **D17** and key in **=D2+(D13*D6)**. Press [**Enter**] and the calculated value of the upper limit, 23.4868, will appear in the output table.

15	6	CONFIDENCE INTERVAL	
16	11	Lower Limit	5.579839
17		Upper Limit	23.48683

Interpreting the Confidence Interval

The confidence interval for a population mean provides the range of values that, with a specified degree of confidence, includes the unknown population mean. When the statistical test result is significant, the sample mean is considered to be a good estimate of the population mean. Therefore, the sample mean is used when computing the upper and lower limits of the confidence interval. Because alpha was set at .01, a 99% confidence interval was constructed. The 99% confidence interval for this TV viewing research problem looks like: $5.5798 \leq \mu \leq 23.4868$.

Based on the limits of this interval, we would claim that we are 99% confident that the true population mean falls between 5.5798 and 23.4868 hours. Expressed another way, if this analysis were carried out repeatedly, sampling all possible samples of 15 Canadian adults, 99% of the confidence intervals would contain the true population mean. Only 1% of the intervals would not contain the true population mean. Because the hypothesized mean value of 28 is not included in the confidence interval, it is consistent with the previous conclusion based on the value of obtained t; namely, the result is statistically significant and the true population mean is a value other than 28.

Chapter 8

Testing Hypotheses About the Difference Between Two Means

Researchers frequently want to investigate the effects of experimental treatments or the performance differences between existing groups such as males and females. The two broad categories of tests where observations are in two groups are: 1) independent, where data have been collected in a manner such that observations are not related to one another and 2) dependent (sometimes called paired or correlated), where either the same person is measured twice under different conditions or pairs of subjects are chosen so that they are similar to each other. In either case, the data are summarized in the form of sample means that can be compared using a t-test or a Z-test. First I present the independent samples t-test, then the dependent samples t-test, and finally the Z-test for two independent samples.

t-Test for Two Independent Samples

The independent samples t-test that is most commonly used is the one that utilizes a pooled variance in the calculation of the standard error of the difference. To justify pooling the variances of two samples, one needs to assume that the variances of the two populations are equal. If this assumption cannot be met, then an alternative test is available that does not use a pooled variance. I first present the independent samples t-test that assumes equal population variances and then the alternative test that assumes unequal population variances.

Variances Are Not Known and Are Assumed to Be Equal

The formula for the independent samples t-test employing a pooled variance is

$$t = \frac{(\overline{X}_1 - \overline{X}_2) - (\mu_1 - \mu_2)}{S_{\overline{X}_1 - \overline{X}_2}}$$

where $(\overline{X}_1 - \overline{X}_2)$ is the difference between the two sample means, $(\mu_1 - \mu_2)$ is the hypothesized difference between the population means, and $S_{\overline{X}_1 - \overline{X}_2}$ is the standard error of the difference.

The standard error is calculated using a pooled variance estimate. The formula for the pooled variance is

$$S^2_{pooled} = \frac{(n_1 - 1)S_1^2 + (n_2 - 1)S_2^2}{(n_1 - 1) + (n_2 - 1)}$$

where S_1^2 is the variance of sample 1, S_2^2 is the variance of sample 2, n_1 is the number of observations in sample 1, and n_2 is the number of observations in sample 2. As shown in this formula, the pooled variance estimate is the weighted average of the sample variances where each variance is weighted by its respective degrees of freedom.

The formula for the standard error of the difference, then, is given by

$$S_{\overline{X}_1 - \overline{X}_2} = \sqrt{\frac{S^2_{pooled}}{n_1} + \frac{S^2_{pooled}}{n_2}}$$

Assumptions Underlying the Independent Samples t-Test

The statistical assumptions underlying the independent samples t-test are

1. Observations are randomly sampled from population 1 and population 2.

2. The sample of observations from population 1 is independent of the sample of observations from population 2.

3. Observations are normally distributed in both population 1 and population 2.

4. The variances of population 1 and population 2 are unknown but are equal.

Sample Research Problem

A researcher wanted to find out if dreaming increased as a result of taking three milligrams of Melatonin before going to sleep each night. Nineteen people were randomly assigned to one of two treatment conditions: Melatonin (n = 10) and placebo (n = 9). The number of dreams recalled each morning were reported and then tallied over a one-week period.

Steps to Follow to Analyze the Sample Research Problem

1. Enter the data in an Excel worksheet as shown below.

	A	B
1	Melatonin	Placebo
2	21	12
3	18	14
4	14	10
5	20	8
6	11	16
7	19	5
8	8	3
9	12	9
10	13	11
11	15	

2. Click on **Tools** in the Menu bar and select **Data Analysis**.

If Data Analysis does not appear as a choice in the Tools menu, you will need to load the Microsoft Excel ToolPak add-in. Follow the procedure on page 7.

---

3. Three different t-tests can be selected. Select **t-Test: Two-Sample Assuming Equal Variances** and click **OK**.

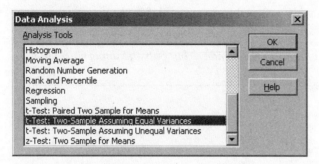

4. Complete the dialog box as shown below. A description of the entries is given immediately after the dialog box.

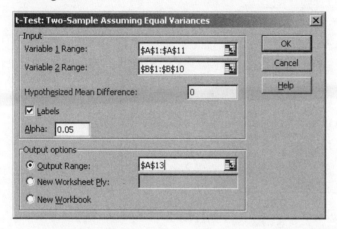

- **Variable 1 Range**. The flashing vertical line should be located in the **Variable 1 Range** window. Click in the top cell of the Melatonin group (cell **A1**) and drag to the end of the data in that column of the worksheet (cell **A11**). Or, if you prefer, manually enter **A1:A11** in the Variable 1 Range window.
- **Variable 2 Range**. Click in the **Variable 2 Range** window. Then click in cell **B1** and drag to the end of the data in that column (cell **B10**). If you prefer, you can manually enter **B1:B10** in the Variable 2 Range window.

  You must click in the Variable 2 Range window before you drag over the range of scores, otherwise the cell addresses will be placed in the Variable 1 Range window.

- **Hypothesized Mean Difference**. This is the hypothesized difference between the population means. This difference is generally set equal to 0, although you could designate a nonzero value if you so desire. If no entry is made, the default value of zero is used. We'll use the default value.
- **Labels**. When checked, this indicates that the first cell in the ranges for sample 1 and sample 2 contains a variable label rather than a sample observation. In this example, these cells contain labels (Melatonin and Placebo). Therefore, click in the **Labels** box to place a check mark there.
- **Alpha**. This is the significance level for the statistical test. Let's use the default value of .05.

- **Output options**. Select the **Output Range option**. This will place the output in the same worksheet as the data. First click in the Output Range window. Then click in cell **A13** of the worksheet to place the output a couple of lines below the data with A13 as the uppermost left cell. If you prefer, you can manually enter **A13** in the window.

  You could indicate the entire range in the Output Range window, but only the uppermost left cell is necessary.

5. Click **OK**.

Interpreting the Output

The output that you will obtain is shown below. If column A is not wide enough to display the output descriptions, you will want to increase the column width.

13	t-Test: Two-Sample Assuming Equal Variances		
14			
15		Melatonin	Placebo
16	Mean	15.1	9.777778
17	Variance	18.32222	16.94444
18	Observations	10	9
19	Pooled Variance	17.67386	
20	Hypothesized Mean Difference	0	
21	df	17	
22	t Stat	2.755318	
23	P(T<=t) one-tail	0.006758	
24	t Critical one-tail	1.739607	
25	P(T<=t) two-tail	0.013516	
26	t Critical two-tail	2.109816	

- **Mean**. Sample means for the two groups, Melatonin ($\overline{X}_1 = 15.1$) and placebo ($\overline{X}_2 = 9.7778$).

- **Variance**. Sample variances for group 1 ($S_1^2 = 18.3222$) and group 2 ($S_2^2 = 16.9444$), calculated using the formula for an unbiased estimate of the population variance.

- **Observations**. The number of observations in group 1 ($n_1 = 10$) and in group 2 ($n_2 = 9$).

- **Pooled Variance**. Pooled variance estimate ($S_{pooled}^2 = 17.6739$), calculated by weighting each group variance by its degrees of freedom.

- **Hypothesized Mean Difference**. The value of 0 indicates that the null hypothesis states that the population means are equal (i.e., $\mu_1 - \mu_2 = 0$).

- **df**. Degrees of freedom for the test (df = 17), calculated by the formula $n_1 + n_2 - 2$.

- **t Stat**. Value of t obtained by applying the formula presented at the beginning of this section. For this research problem, obtained t = 2.7553.

- **P(T<=t) one-tail**. One-tailed chance probability of the obtained t statistic. This probability would be interpreted as follows. If the null hypothesis is true, then the chance probability of obtaining a $t \leq -2.7553$

is .0068. Similarly, the chance probability of obtaining a $t \geq +2.5733$ is .0068. The result is statistically significant if the one-tailed probability is less than alpha.

- **t Critical one-tail**. One-tailed critical t value for selected alpha and df as indicated in the output. For this problem, the one-tailed critical value of t with df equal to 17 and alpha set at .05 is 1.7397. The result is statistically significant if the absolute value of obtained t is greater than critical t.

> *When you compare t-values to determine the statistical significance of a one-tailed test, you must check to be sure that the sign (positive or negative) of the obtained t value is consistent with your alternative hypothesis.*

- **P(T<=t) two-tail**. Two-tailed chance probability of the obtained t statistic. If the null hypothesis is true, then the chance probability of obtaining a $t \leq -2.7553$ plus the chance probability of obtaining a $t \geq +2.7553$ is .0135. The result is statistically significant if the two-tailed probability is less than alpha.

- **t Critical two-tail**. Two-tailed critical t value for the selected alpha and df as indicated in the output. With df equal to 17 and alpha set at .05, the two-tailed critical value of t is 2.1098. The result is statistically significant if the absolute value of the obtained t is greater than critical t.

Variances Are Not Known and Are Assumed to Be Unequal

When the variances of the two populations are both unknown and assumed to be unequal, it may not be appropriate to use a pooled variance estimate. For the situation of unknown and unequal population variances, Excel employs the Welch-Aspin procedure, where the two sample variances, S_1^2 and S_2^2 are used in the calculation of the standard error of the difference. The obtained statistic is symbolized as t*, for which degrees of freedom are not known but are approximated. The formula for t* is given by

$$t^* = \frac{\left(\overline{X}_1 - \overline{X}_2\right) - \left(\mu_1 - \mu_2\right)}{S_{\overline{X}_1 - \overline{X}_2}}$$

which is the same as the test assuming equal variances, except for the calculation of the standard error of the difference. The standard error formula is

$$S_{\overline{X}_1 - \overline{X}_2} = \sqrt{\frac{S_1^2}{n_1} + \frac{S_2^2}{n_2}}$$

The degrees of freedom are approximated by the following formula:

$$df^* = \frac{df_1 df_2}{df_2 C^2 + df_1 (1 - C)^2}$$

The C term in this formula is found by

$$C = \frac{\left(\dfrac{S_1^2}{n_1}\right)}{\left(\dfrac{S_1^2}{n_1}\right) + \left(\dfrac{S_2^2}{n_2}\right)}$$

Assumptions Underlying the Independent Samples t*-Test

The assumptions underlying the Welch-Aspin t*-test are the same as those underlying the independent samples t-test, with the exception of the equality of population variances assumption. The Welch-Aspin t* is based on the assumption that the population variances are not known and are not equal. In practice, the Welch-Aspin procedure is infrequently used, probably because of two considerations. First, when the sample sizes, n_1 and n_2, are equal, the pooled variances t-test is robust with respect to violations of the homogeneous variances assumption. Second, when sample sizes are both relatively large, say, greater than 30, the pooled variances t provides a satisfactory approximation. Thus, it is only when sample sizes are small and markedly unequal that t* would be strongly recommended for analyzing data with unequal population variances.

Sample Research Problem

An animal researcher carried out an experiment designed to find out if the type of food reward given to rats for successfully negotiating a maze would have an effect on their maze-running times. Four rats were assigned to receive bran, an especially attractive food, and eight different rats were assigned to receive their regular food as a reward. The running times in seconds were recorded after 50 trials

Steps to Follow to Analyze the Sample Research Problem

1. Enter the data in an Excel worksheet as shown below.

	A	B
1	Bran	Regular
2	32	118
3	38	78
4	27	82
5	45	91
6		67
7		41
8		97
9		42

2. Click on **Tools** in the Menu Bar and then select **Data Analysis**.

If Data Analysis does not appear as a choice in the Tools menu, you will need to load the Microsoft Excel ToolPak add-in. Follow the procedure on page 7.

---

3. In the Data Analysis dialog box, select **t-Test: Two-Sample Assuming Unequal Variances** and click **OK**.

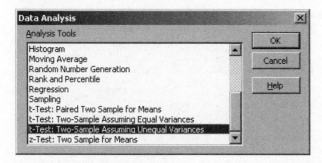

4. Complete the dialog box as shown below. Detailed instructions regarding the entries are given immediately following the dialog box.

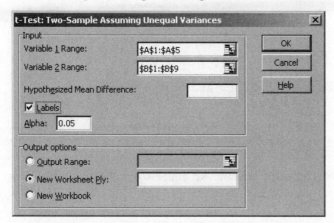

- **Variable 1 Range**. Enter the worksheet location of the Bran condition. To do this, first click in the **Variable 1 Range** window. Then click in cell **A1** of the worksheet and drag to cell **A5**. If you prefer, you can manually enter **A1:A5**.

- **Variable 2 Range**. Enter the worksheet location of the Regular condition. To do this, first click in the **Variable 2 Range** window. Then click in cell **B1** of the worksheet and drag to cell **B9**. If you prefer, you can manually enter **B1:B9**.

- **Hypothesized Mean Difference**. We'll use the default value of zero. You do not have to enter anything here if you are using the default value.

- **Labels**. A check mark should appear in the box to the left of **Labels**, because the top cell in each of the variable ranges contains a label (Bran or Regular) rather than a data value.

- **Alpha**. Let's use the default value of .05 for the significance level of the statistical test.

- **Output options**. Let's place the output in a new worksheet. To do this, click the button to the left of **New Worksheet Ply** so that a black dot appears there.

5. Click **OK**.

Interpreting the Output

The output that you will obtain for the maze-running experiment is displayed below. I increased the width of column A so that I could read each label in its entirety.

	A	B	C
1	t-Test: Two-Sample Assuming Unequal Variances		
2			
3		Bran	Regular
4	Mean	35.5	77
5	Variance	60.33333	703.4286
6	Observations	4	8
7	Hypothesized Mean Difference	0	
8	df	9	
9	t Stat	-4.08888	
10	P(T<=t) one-tail	0.001361	
11	t Critical one-tail	1.833113	
12	P(T<=t) two-tail	0.002722	
13	t Critical two-tail	2.262157	

- **Mean**. Sample mean for the bran condition ($\overline{X}_1 = 35.5$) and sample mean for the regular condition ($\overline{X}_2 = 77$).

- **Variance**. Sample variance for the bran condition ($S_1^2 = 60.3333$) and the regular condition ($S_2^2 = 703.4286$).

- **Observations**. Number of observations in the bran condition ($n_1 = 4$) and the regular condition ($n_2 = 8$).

- **Hypothesized Mean Difference**. The hypothesized difference was set equal to 0, indicating that the hypothesis under test was that $\mu_1 - \mu_2 = 0$.

- **df**. Degrees of freedom for the t*-test (df = 9), calculated via the Welch-Aspin procedure.

- **t Stat**. Value of t* obtained by applying the Welch-Aspin procedure. For this research problem, t* = – 4.0889.

- **P(T<=t) one-tail**. One-tailed chance probability of the obtained t*. If the null hypothesis is assumed to be true, then the chance probability of obtaining a t* ≤ – 4.0889 is .0014. Similarly, .0014 is the chance probability associated with the obtained test statistic value in the upper tail of the distribution (i.e., t* ≥ +4.0889). A one-tailed test is statistically significant if the one-tailed probability is less than alpha.

- **t Critical one-tail**. One-tailed critical t* value for designated alpha and df as indicated in the output. For this research problem, the one-tailed critical t* value of 1.8331 is for alpha equal to .05 and 9 df. A one-tailed test is statistically significant if the absolute value of obtained t is greater than critical t.

When you compare t-values to determine the statistical significance of a one-tailed test, you must check to be sure that the sign (positive or negative) of the obtained t value is consistent with your alternative hypothesis.

- **P(T<=t) two-tail**. The two-tailed chance probability of obtained t* equals .0027. It includes both the chance probability that t* ≤ -4.0889 and t* $\geq +4.0889$, or double the one-tailed probability. A two-tailed test is statistically significant if the two-tailed probability is less than alpha.

- **t Critical two-tail**. Two-tailed critical t* value for designated alpha and df as calculated according to the Welch-Aspin procedure. With df equal to 9 and alpha set at .05, the two-tailed critical value is 2.2622. A two-tailed test is statistically significant if the absolute value of obtained t is greater than critical t.

Paired-Samples t-Test

The paired-samples t-test is commonly used for two types of data analysis situations. The first is what is generally referred to as a before-after design, where the same research subjects are measured before and after an experimental treatment intervention. The second involves matching pairs of research subjects on a variable that is believed to be correlated with the dependent variable. For example, to analyze the data from an experiment on weight loss, persons could be matched on initial weight. Or, in a study on the effectiveness of instructional programs, students could be matched on intelligence test scores. The paired-samples t-test is probably more widely known to researchers as the correlated- or dependent-samples t-test. The problem that I use to illustrate the paired-samples t test utilizes the data collected in a study that employed a before-after design.

The formula for the matched-samples t-test is given by

$$t = \frac{\overline{D} - \mu_D}{S_{\overline{D}}}$$

where $\overline{D}$ is the sample mean difference score, μ_D is the hypothesized population mean difference, and $S_{\overline{D}}$ is the standard error of the mean difference.

The standard error of the mean difference is found by the following formula:

$$S_{\overline{D}} = \frac{S_D}{\sqrt{n}}$$

where S_D is the sample standard deviation of the difference scores, and n is the total number of pairs of observations.

Finally, the formula for S_D is given by

$$S_D = \sqrt{\frac{\sum (D - \overline{D})^2}{n - 1}}$$

where D is a before-after difference score $(X_1 - X_2)$ and $\overline{D}$ is the mean difference score.

Assumptions Underlying the Paired-Samples t-Test

There are four assumptions underlying the paired samples t-test.

1. Observations are randomly sampled from the population(s) of interest.

2. The observations are correlated.

3. Observations are normally distributed in the population(s).

4. The variance of the difference scores is unknown.

Sample Research Problem

A researcher was concerned about the debilitating effects of test anxiety on academic performance and conducted a study to investigate the effectiveness of relaxation training for reducing anxiety. Six students who had been identified as having high test anxiety were randomly selected to participate in a relaxation training program. Before training began, each student completed a test developed to measure test anxiety. After the training was completed, each student again completed the same anxiety test. The scores obtained on the two administrations were recorded in an Excel worksheet

Steps to Follow to Analyze the Sample Research Problem

1. Enter the data in an Excel worksheet as shown below.

	A	B
1	Before	After
2	84	72
3	76	70
4	104	90
5	103	94
6	91	93
7	90	90

2. Click on **Tools** in the Menu Bar and then select **Data Analysis**.

If Data Analysis does not appear as a choice in the Tools menu, you will need to load the Microsoft Excel ToolPak addin. Follow the procedure on page 7.

3. In the Data Analysis dialog box, select **t-Test: Paired Two Sample for Means** click **OK**.

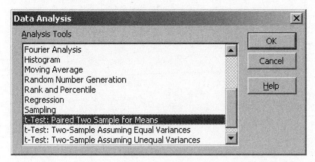

4. Complete the dialog box as shown below. Detailed information about the entries is given immediately following the dialog box.

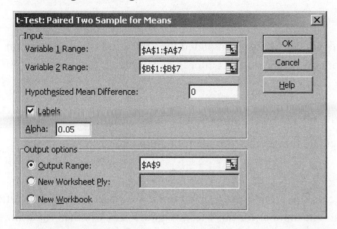

- **Variable 1 Range**. Enter the worksheet location of the Before condition. To do this, click in the **Variable 1 Range** window. Then, click in cell **A1** in the worksheet and drag to cell **A7**. If you prefer, you can manually enter **A1:A7**.
- **Variable 2 Range**. Enter the worksheet location of the After condition. Click in the **Variable 2 Range** window. Then click in cell **B1** of the worksheet and drag to cell **B7**. If you prefer, you can manually enter **B1:B7**.
- **Hypothesized Mean Difference**. Enter **0** to test the hypothesis that the population mean difference is equal to zero.
- **Labels**. The top cell in each of the variable ranges contains a label (Before and After), so a check mark should appear in the **Labels** box. Click in the box to place a check mark there.
- **Alpha**. Alpha refers to the significance level for the statistical test. Let's use the default value of .05.
- **Output options**. Let's place the output in the same worksheet as the data. Click in the **Output Range** window. Then click on cell **A9** in the worksheet. Cell A9 will be the uppermost left cell of the output. If you prefer, you can manually enter **A9** in the Output Range window.

5. Click **OK**.

Interpreting the Output

The output that you will obtain for the paired-samples t-test example is shown below.

9	t-Test: Paired Two Sample for Means		
10			
11		Before	After
12	Mean	91.33333	84.83333
13	Variance	117.4667	117.7667
14	Observations	6	6
15	Pearson Correlation	0.82358	
16	Hypothesized Mean Difference	0	
17	df	5	
18	t Stat	2.471525	
19	P(T<=t) one-tail	0.028212	
20	t Critical one-tail	2.015048	
21	P(T<=t) two-tail	0.056425	
22	t Critical two-tail	2.570582	

- **Mean**. Mean anxiety test score of the six students before they completed the relaxation training program ($\overline{X}_1 = 91.3333$) and the mean anxiety test score of the six students after they completed the program ($\overline{X}_2 = 84.8333$).

- **Variance**. Variance of the test taken before the relaxation training ($S_1^2 = 117.4667$) and variance of the test taken after training ($S_2^2 = 117.7667$), both calculated using the formula for an unbiased estimate of the population variance.

- **Observations**. Number of observations in the before group ($n_1 = 6$) and number of observations in the after group ($n_2 = 6$). The n's will always be equal to one another for this t-test, regardless of whether you utilized a before-after design or a matched-pairs design.

- **Pearson Correlation**. The Pearson correlation coefficient, expressing the strength of the relationship between the pairs of before-after observations. The correlation of .8236 indicates a strong relationship between the pairs of observations.

- **Hypothesized Mean Difference**. This was designated as 0 in the dialog box, implying that the hypothesis under test was $\mu_1 - \mu_2 = 0$, or, alternatively, that the population mean difference (μ_D) was equal to 0.

- **df**. Degrees of freedom for the test (df = 5), calculated by the formula n − 1, where n refers to the total number of pairs of observations.

- **t Stat**. Value of t obtained by applying the formula for the paired-samples t-test. For the example analysis, obtained t = 2.4715.

- **P(T<=t) one-tail**. One-tailed chance probability of the obtained t statistic. More specifically, if the null hypothesis is true, then the chance probability of obtaining a t ≤ −2.4715 is .0282. Similarly, the chance probability of obtaining a t ≥ +2.4715 is .0282. A one-tailed test is statistically significant if the one-tailed probability is less than alpha.

- **t Critical one-tail**. One-tailed critical t value for designated alpha (.05) and df equal to 5. For this analysis, the one-tailed critical t is equal to 2.0150. A one-tailed test is statistically significant if the absolute value of the obtained t is greater than critical t.

> *When you compare t-values to determine the statistical significance of a one-tailed test, you must check to be sure that the sign (positive or negative) of the obtained t value is consistent with your alternative hypothesis.*

- **P(T<=t) two-tail**. Two-tailed chance probability of the obtained t statistic. That is, if the null hypothesis is true, then the chance probability of obtaining a t $\leq$ −2.4715 plus the chance probability of obtaining a t $\geq$ +2.4715 is .0564. A two-tailed test is statistically significant if the two-tailed probability is less than alpha.

- **t Critical two-tail**. Two-tailed critical value for designated alpha (.05) and df equal to 5. In this case, the critical two-tailed t is equal to 2.5706. A two-tailed test is statistically significant if obtained t is greater than the two-tailed critical t.

Z-Test for Two Independent Samples

The formula for the independent samples Z-test is

$$Z = \frac{(\overline{X}_1 - \overline{X}_2) - (\mu_1 - \mu_2)}{\sigma_{\overline{X}_1 - \overline{X}_2}}$$

where $(\overline{X}_1 - \overline{X}_2)$ is the difference between the two sample means, $(\mu_1 - \mu_2)$ is the hypothesized difference between the population means, and $\sigma_{\overline{X}_1 - \overline{X}_2}$ is the standard error of the difference.

The formula for the standard error of the difference is given by

$$\sigma_{\overline{X}_1 - \overline{X}_2} = \sqrt{\frac{\sigma_1^2}{n_1} + \frac{\sigma_2^2}{n_2}}$$

where σ_1^2 is the variance of population 1, σ_2^2 is the variance of population 2, n_1 is the size of sample 1, and n_2 is the size of sample 2.

Assumptions Underlying the Z-Test

The assumptions underlying the two-samples Z-test are similar to those for the independent samples t-test. An important difference, however, is that the Z-test assumes that population variances are known. This assumption limits the usefulness of the Z-test, because population parameters are seldom known and must be estimated from the sample data.

The statistical assumptions underlying the two-sample Z-test include:

1. Observations are randomly sampled from population 1 and population 2.

2. The samples from the two populations are independent of each other.

3. Observations are normally distributed in population 1 and population 2.

4. The variances of the two populations, σ_1^2 and σ_2^2, are known.

Sample Research Problem

The mathematics performance of boys is generally believed to be higher than that of girls, especially at more advanced levels of mathematics. A researcher was interested in determining if this mathematics performance difference appeared as early as second grade. Nine second-grade boys and nine second-grade girls were randomly selected and administered a standardized mathematics achievement test for which it is known that the population variance for boys is 256 and the population variance for girls is 324.

Steps to Follow to Analyze the Sample Research Problem

1. Enter the data in an Excel worksheet as shown below.

	A	B
1	Boys	Girls
2	45	51
3	34	36
4	51	24
5	46	43
6	51	37
7	31	39
8	48	52
9	25	37
10	50	48

2. Click on **Tools** in the Menu Bar and select **Data Analysis**.

If Data Analysis does not appear as a choice in the Tools menu, you will need to load the Microsoft Excel ToolPak add-in. Follow the procedure on page 7.

3. In the Data Analysis dialog box, select **z-Test: Two Sample for Means** and click **OK**.

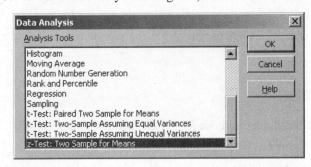

4. Complete the dialog box as shown below. A description of each entry is given immediately after the dialog box.

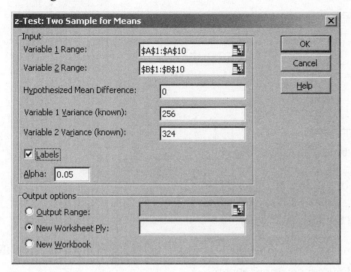

- **Variable 1 Range.** The flashing vertical line should be located in the **Variable 1 Range** window. Click in the top cell of the Boys' data (cell **A1**) and drag to the end of the data in that column (**A10**). If you prefer, you can manually enter **A1:A10** in the Variable 1 Range window.
- **Variable 2 Range.** Click in the **Variable 2 Range** window. Then click in cell **B1** in the worksheet and drag to the end of the data in that column (**B10**). Or, if you prefer, you can manually enter **B1:B10** in the Variable 2 Range window.

  You must click in the Variable 2 Range box before you drag over the range of scores, otherwise the addresses will be placed in the wrong location.

- **Hypothesized Mean Difference.** Enter a **0** for the value of the Hypothesized Mean Difference.
- **Variable 1 Variance (known).** Enter **256** for the boys' population variance.
- **Variable 2 Variance (known).** Enter **324** for the girls' population variance.
- **Labels.** Because group names (Boys and Girls) were included in the variable range, the **Labels** box must be checked. Click in the box to place a check mark there.
- **Alpha.** Let's use .05 for alpha, the default value.
- **Output options.** Let's place the output in a new worksheet. To do this, click in the button to the left of **New Worksheet Ply** so that a black dot appears there.

5. Click **OK**.

Interpreting the Output

The output for this analysis is displayed at the top of the next page. If column A is not wide enough to display the output descriptions, you may want to increase the width.

	A	B	C
1	z-Test: Two Sample for Means		
2			
3		Boys	Girls
4	Mean	42.33333	40.77778
5	Known Variance	256	324
6	Observations	9	9
7	Hypothesized Mean Difference	0	
8	z	0.193773	
9	P(Z<=z) one-tail	0.423177	
10	z Critical one-tail	1.644854	
11	P(Z<=z) two-tail	0.846354	
12	z Critical two-tail	1.959964	

- **Mean**. Sample means for the two groups, boys ($\overline{X}_1 = 42.3333$) and girls ($\overline{X}_2 = 40.7778$).

- **Known Variance**. Population variances, $\sigma_1^2 = 256$ and $\sigma_2^2 = 324$.

- **Observations**. Number of observations in group 1 ($n_1 = 9$) and in group 2 ($n_2 = 9$).

- **Hypothesized Mean Difference**. The value 0 indicates that the hypothesis under test in this analysis is that the population means of the second-grade boys and girls are equal.

- **z**. Value of Z obtained by applying the formula presented at the beginning of this section. For this problem, obtained Z = 0.1938.

- **P(Z<=z) one-tail**. One-tailed chance probability of the obtained Z. That is, if the null hypothesis is true, then the chance probability of obtaining a Z $\leq$ –0.1938 is .4232. Or, in the upper tail of the distribution, the chance probability of obtaining a Z $\geq$ +0.1938 is .4232. The one-tailed test is statistically significant if the one-tailed probability is less than alpha.

- **z Critical one-tail**. One-tailed critical value of Z for the alpha selected in the dialog box. In this case, with alpha set equal to .05, the one-tailed critical value of Z is 1.6449. The one-tailed test is statistically significant if the absolute value of obtained Z is greater than the one-tailed critical value.

> *When you compare Z-values to determine the statistical significance of a one-tailed test, you must check to be sure that the sign (positive or negative) of the obtained Z value is consistent with your alternative hypothesis.*

- **P(Z<=z) two-tail**. Two-tailed chance probability of obtained Z. If the null hypothesis is true, then the chance probability of obtaining a Z $\leq$ –0.1938 plus the chance probability of obtaining a Z $\geq$ +0.1938 is two times the one-tailed probability, 2 X .4232 = .8464. The two-tailed test is statistically significant if the two-tailed probability is less than alpha.

- **z Critical two-tail**. Two-tailed critical value of Z for the alpha selected in the dialog box. With alpha set at .05, the two-tailed critical value is 1.9600. The two-tailed test is statistically significant if the absolute value of obtained Z is greater than the two-tailed critical value.

Research designs often include more than one independent variable, as well as independent variables that have more than two levels. If a researcher is using such designs and is interested in differences among group means, analysis of variance (ANOVA) may be the appropriate statistical technique. Excel provides Data Analysis Tools for carrying out a one-way between-groups ANOVA, a one-way repeated measures ANOVA, and a two-way between-groups ANOVA.

One-Way Between-Groups ANOVA

F-Test

To carry out a one-way between-groups ANOVA, one applies the F-test, which, for this analysis, is the ratio of mean square (MS) between to MS within:

$$F = \frac{MS_{Between}}{MS_{Within}}$$

Assumptions Underlying the F -Test

The statistical assumptions underlying the F-test include

1. Samples are randomly selected from the K populations.

2. The observations from the K populations are independent of one another.

3. Observations are normally distributed in the K populations.

4. The variances of the K populations are not known but are equal to one another.

Sample Research Problem

Sixteen adults were assigned to one of three conditions in which they were given a sheet with background information (e.g., sex, age, employment history) on a defendant who was being charged with child abuse. The three conditions were

1. **Clean record**. The information sheet stated that the defendant had a completely clean criminal record.

2. **Criminal record**. The information sheet stated that the defendant had a previous record of child abuse.

3. **Control**. The information sheet gave no information about the defendant's criminal record.

Adults in all three conditions were asked to view a video of the trial and then to rate their perception of the defendant's guilt or innocence. They were asked to utilize a 7-point rating scale, where 1 referred to "completely sure the defendant is guilty," and 7 referred to "completely sure the defendant is innocent." The researcher was interested in finding out if knowledge of a previous record would influence individuals' perceptions of guilt or innocence.

Using Analysis Tools for One-Way Between-Groups ANOVA

1. Enter the participants' ratings in an Excel worksheet as shown below.

	A	B	C
1	Clean	Criminal	Control
2	6	1	4
3	5	2	3
4	7	1	5
5	6	3	6
6	3	2	4
7	7		

2. Click on **Tools** in the Menu Bar and select **Data Analysis**.

   If Data Analysis does not appear as a choice in the Tools menu, you will need to load the Microsoft Excel ToolPak add-in. Follow the procedure on page 7.

3. In the Data Analysis dialog box, select the Analysis Tool named **Anova: Single Factor**, and then click **OK**.

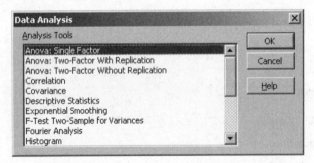

4. Complete the Anova dialog box as shown below. Detailed information regarding the entries is given immediately after the dialog box.

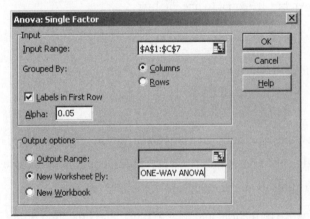

- **Input Range**. Enter the worksheet location of the data you wish to analyze. Click in the **Input Range** window. Then click in cell **A1** and drag to the end of the data in cell **C7**. If you prefer, you could also key in the data range by entering **A1:C7**. This input range includes the condition labels: Clean, Criminal, and Control.

- **Grouped by**. The data are grouped by columns in the worksheet. **Columns** was automatically selected, so you do not need to do anything additional for this step.

- **Labels in First Row**. The top cell in each column (A1, B1, and C1) of the data range contains a treatment condition label. To let Excel know that these are labels and should not be included in the data analysis, click in the box for **Labels in First Row.**

- **Alpha**. The default value of alpha is .05. If you want to carry out the test with a different alpha, just enter the desired significance level in the space to the right of **Alpha**. We will use the default value for this example.

- **Output options**. Three output options are available. Let's use the default option of **New Worksheet Ply**. The output table will be pasted into a new worksheet with A1 as the uppermost left cell. If you utilize the new worksheet option, you may give the worksheet a name that will enable you to identify it easily. Let's name the new worksheet **ONE-WAY ANOVA**.

5. Click **OK**. The output will be displayed in the ONE-WAY ANOVA worksheet.

	A	B	C	D	E	F	G
1	Anova: Single Factor						
2							
3	SUMMARY						
4	*Groups*	*Count*	*Sum*	*Average*	*Variance*		
5	Clean	6	34	5.666667	2.266667		
6	Criminal	5	9	1.8	0.7		
7	Control	5	22	4.4	1.3		
8							
9							
10	ANOVA						
11	*Source of Variation*	*SS*	*df*	*MS*	*F*	*P-value*	*F crit*
12	Between Groups	41.60417	2	20.80208	13.98761	0.000574	3.805565
13	Within Groups	19.33333	13	1.487179			
14							
15	Total	60.9375	15				

Interpreting the Output

The output provided for Anova: Single Factor (shown above) includes summary information about the groups and an ANOVA table.

SUMMARY

- **Groups**. The label for each of the three groups: Clean, Criminal, and Control. If the row with labels had not been included in the input range, the three groups would be identified as: Column 1, Column 2, and Column 3.

- **Count**. The number of observations in each group.

- **Sum**. The sum of the guilt/innocence ratings for each group.

- **Average**. The mean of the guilt/innocence ratings for each group.

- **Variance**. The variance of the guilt/innocence ratings for each group.

ANOVA Table

- **Source of Variation**. In a one-way between-groups ANOVA, the total variation is partitioned into two sources, between groups and within groups. Between groups refers to the degree to which the three groups differ from each other with respect to their guilt/innocence ratings. Within groups refers to the degree to which the adults within each group differ from one another.

- **SS**. Abbreviation for **sum of squares**.

    $SS_{Between}$ is obtained by applying the formula

    $$SS_{Between} = \sum_{K=1}^{K} \left(\overline{X}_K - \overline{X}_G \right)^2$$

    where $\overline{X}_K$ is the mean of the kth group, and

    $\overline{X}_G$ is the grand mean across all subjects.

    SS_{Within} is obtained by applying the formula

    $$SS_{Within} = \sum_{K=1}^{K} (n_K - 1)S_K^2$$

    where n_K is the number of subjects in the kth group, and

    S_K^2 is the variance of the kth group.

- **df**. Abbreviation for degrees of freedom. $df_{Between} = K–1$ and $df_{Within} = N–K$, where N refers to the total number of subjects.

- **MS**. Abbreviation for mean square. $MS_{Between}$ is computed by dividing $SS_{Between}$ by $df_{Between}$. Similarly, MS_{Within} is computed by dividing SS_{Within} by df_{Within}.

- **F**. The obtained value of the F test statistic, found by applying the formula

    $$F = \frac{MS_{Between}}{MS_{Within}}$$

    For the sample research problem,

    $$F = \frac{20.8021}{1.4872} = 13.9876$$

- **P-value**. The chance probability associated with the obtained value of F. If the null hypothesis is true that the population means for the three groups are all equal to one another, then the chance probability of obtaining an F $\geq$ 13.9876 is .0006. Because the P-value is less than alpha (.05), we would conclude that the result is statistically significant and that the three population means are not all equal to one another.

- **F crit**. The critical value of F associated with the selected value of alpha (.05). Because the obtained F of 13.9876 is greater than the critical F of 3.8056, we would conclude that the result is statistically significant and that the three population means are not all equal to one another. Note that the conclusion reached by comparing the P-value to alpha should be the same conclusion that is reached by comparing the critical value of F to the obtained value of F.

One-Way Repeated Measures ANOVA

F-Test

The one-way between-groups ANOVA and the one-way repeated measures ANOVA differ with respect to how many times the subjects are measured or tested. Each subject is measured only once in the between-groups model. In the repeated measures model, each subject is measured three or more times. If subjects are measured just two times, the paired-samples t-test should be utilized. The F-test for the repeated measures model is given by

$$F = \frac{MS_{Treatments}}{MS_{Error}}$$

Assumptions Underlying the F-Test

1. The sample is randomly selected from the population of interest.
2. Each population of observations is normally distributed.
3. The variances of the K populations are not known but are equal to one another.
4. The population covariances for all pairs of treatment levels are equal.

Sample Research Problem

An investigator was interested in assessing the effects of drugs on driving skills. Six subjects performed the same driving simulation task under three different treatment conditions: 1) while under the influence of marijuana, 2) while under the influence of alcohol, and 3) while under the influence of no drugs. Driving simulation scores could range from 0 to 35, where higher scores were associated with better performance. The order of treatment administration was counterbalanced across the subjects. The data were recorded in an Excel worksheet with subject numbers in the first column and simulation scores in the next three columns.

Using Analysis Tools for One-Way Repeated Measures ANOVA

1. Enter the data in an Excel worksheet as shown below.

	A	B	C	D
1	Subject	Marijuana	Alcohol	No Drug
2	1	19	8	21
3	2	18	10	31
4	3	25	10	26
5	4	20	18	28
6	5	17	7	14
7	6	21	16	24

2. Click on **Tools** in the Menu Bar and select **Data Analysis**.

> *If Data Analysis does not appear as a choice in the Tools menu, you will need to load the Microsoft Excel ToolPak add-in. Follow the procedure on page 7.*

3. In the Data Analysis dialog box, select **Anova: Two-Factor Without Replication** and click **OK**.

> *"Anova: Two-Factor **Without Replication**" seems to be an unusual name for the one-way **repeated** measures ANOVA. The name, however, is appropriately descriptive. The two factors are subjects and treatments. "Without Replication" refers to n=1 subject per cell. Note that this is also the procedure that would be utilized to perform a randomized blocks ANOVA, in which the two factors are blocks and treatments.*

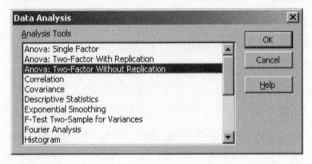

4. Complete the ANOVA dialog box as shown below. Detailed instructions are given immediately after the dialog box.

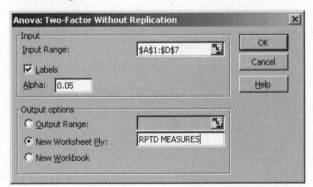

- **Input Range**. Enter the worksheet location of the data you wish to analyze. Click in cell **A1** and drag to the end of the data in **D7**. You could also key in the data range by entering **A1:D7** or **$A$1:$D$7**.

  It is essential that the first column contain identifying information for the subjects (ID numbers or names). If this column is not included in the data range, the output will be inaccurate.

- **Labels**. Click in the box to the left of **Labels** to indicate that the first cell in each of the three treatment columns is a label (i.e., Subject, Marijuana, Alcohol, No Drug) and not a value to be included in the analysis.
- **Alpha**. The default value for alpha is .05. Let's use the default value for this example.
- **Output options**. Select the default option, **New Worksheet Ply**. The output table will be pasted into a new worksheet with A1 as the uppermost left cell. If you select this option, you may give the worksheet an easily identifiable name. Let's name the worksheet **RPTD MEASURES**.

5. Click **OK**. The analysis output will be displayed in the RPTD MEASURES worksheet.

Interpreting the Output

The output provided for the one-way repeated measures ANOVA includes summary information about the subjects, summary information about the treatment conditions, and an ANOVA table. The output is shown below.

	A	B	C	D	E	F	G
1	Anova: Two-Factor Without Replication						
2							
3	SUMMARY	Count	Sum	Average	Variance		
4	1	3	48	16	49		
5	2	3	59	19.66667	112.3333		
6	3	3	61	20.33333	80.33333		
7	4	3	66	22	28		
8	5	3	38	12.66667	26.33333		
9	6	3	61	20.33333	16.33333		
10							
11	Marijuana	6	120	20	8		
12	Alcohol	6	69	11.5	19.9		
13	No Drug	6	144	24	35.6		
14							
15							
16	ANOVA						
17	Source of Variation	SS	df	MS	F	P-value	F crit
18	Rows	181.8333	5	36.36667	2.68059	0.086618	3.325835
19	Columns	489	2	244.5	18.02211	0.000483	4.102821
20	Error	135.6667	10	13.56667			
21							
22	Total	806.5	17				

SUMMARY

- **Count**. The total number of scores recorded for each subject, and, at the bottom of the summary section, the total number of scores recorded for each treatment condition. For this investigation, six subjects were tested three times each.

- **Sum**. The sum of each subject's scores across the three treatment conditions, and, at the bottom, the sum of the scores in each condition.

- **Average**. Each subject's mean score, and, at the bottom, the mean score obtained by the six subjects in each condition.

- **Variance**. The variance of each subject's scores, and, at the bottom, the variance of the six scores in each condition.

ANOVA Table

- **Source of Variation**

  Rows. Each subject's scores were recorded in a separate row of the worksheet, making subjects the row factor for this problem.

  Columns. Each treatment occupied a separate column in the worksheet, making treatments the column factor for this problem.

  Error. Error is the subjects X treatments interaction.

- **SS**

  Sum of squares total in the repeated measures model is partitioned into $SS_{Subjects}$, $SS_{Treatments}$, and $SS_{Interaction}$. For clarity in presentation of the formulas, I first computed I, II, III, IV, and V as shown below, and then utilized these values to compute each SS.

$$I = \sum_{N=1}^{N} X_N^2 \text{, where N refers to the total number of observations (N=18).}$$

$$I = 19^2 + 8^2 + 21^2 + \ldots + 24^2 = 6967$$

$$II = \frac{\left(\sum_{N=1}^{N} X\right)^2}{N} \text{, where N again refers to the total numbers of observations.}$$

$$II = \frac{(333)^2}{18} = 6160.5$$

$$III = \frac{\sum_{K=1}^{K} T^2}{n} \text{, where T is the sum of the scores in each treatment condition, K is the number of treatments,}$$

and n is the number of subjects in each level of the treatment variable.

$$III = \frac{(120)^2}{6} + \frac{(69)^2}{6} + \frac{(144)^2}{6} = 6649.5$$

$$IV = \dfrac{\sum\limits_{n=1}^{n} S^2}{K}$$, where S is the sum of the scores for each subject.

$$IV = \dfrac{(48)^2}{3} + \dfrac{(59)^2}{3} + \dfrac{(61)^2}{3} + ... + \dfrac{(61)^2}{3} = 6342.3333$$

$$SS_{Subjects} = IV - II = 181.8333$$

$$SS_{Treatments} = III - II = 489$$

$$SS_{Error} = I + II - III - IV = 135.6667$$

$$SS_{Total} = I - II = 806.5$$

- **df**. The formulas for df are as follows:

$$df_{Subjects} = (n - 1),$$ where n is the number of subjects.

$$df_{Subjects} = (6 - 1) = 5$$

$$df_{Treatments} = (K - 1),$$ where K is the number of treatments.

$$df_{Treatments} = (3 - 1) = 2$$

$$df_{Error} = (n - 1)(K - 1)$$

$$df_{Error} = (6 - 1)(3 - 1) = 10$$

$$df_{Total} = nK - 1$$

$$df_{Total} = (6)(3) - 1 = 17$$

- **MS**. The mean square terms are found by dividing each SS by its respective df.

- **F**. The obtained F associated with the subjects factor typically is not evaluated. The F associated with the treatments factor (columns) is found by dividing $MS_{Treatments}$ by MS_{Error}. Here, $F_{Treatments} = 18.0221$.

- **P-value**. Chance probability of obtained F if the null hypothesis is true that the three treatment means are equal to one another in the population. Because the P-value of .0005 is less than .05 (alpha), we would conclude that the result is statistically significant.

- **F crit**. The critical value of F distributed with $df_{Treatments} = 2$ and $df_{Error} = 10$, and alpha set at .05. Because the obtained F of 18.0221 exceeds the critical F of 4.1028, we would conclude that the result is statistically significant.

Two-Way Between-Groups ANOVA

A two-way ANOVA allows a researcher simultaneously to analyze two factors and the interaction between the two factors. The procedure provided in Data Analysis Tools requires that both factors are between groups. That is to say, each subject can be included in only one group. Further, the cell n's must be equal.

F-Test

Three different F-tests are associated with the two-way between-groups ANOVA. The F for the main effect of factor A is given by

$$F_A = \frac{MS_A}{MS_{Within}}$$

The F for the main effect of factor B is given by

$$F_B = \frac{MS_B}{MS_{Within}}$$

Finally, the F for the A x B interaction is given by

$$F_{AXB} = \frac{MS_{AXB}}{MS_{Within}}$$

Assumptions Underlying the F -Test

The same assumptions apply to each of the three F-tests.

1. Subjects are randomly sampled from the populations of interest.

2. Observations are distributed normally in each population representing the combination of a level of A with a level of B.

3. Population variances are equal across all combinations of a level of A with a level of B.

4. Observations are independent, both within and across treatment combinations.

Sample Research Problem

An educational researcher was interested in the mathematics performance of low-achieving middle-school children. Although it has been found that performance benefits are associated with small-group work in mathematics, the researcher wanted to find out if the sex composition of the small groups made a difference. Of specific interest were dyads, in which each low-achieving student worked along with one high-achieving student. Would it be more beneficial for low achievers if they worked with high-achieving students whose sex was the same as their own? The researcher randomly selected 15 boys and 15 girls who were low achievers in mathematics and assigned them in equal numbers to one of three treatment conditions to work on a set of mathematics problems: 1) work with one high-achieving student of the same sex, 2) work with one high-achieving student of the opposite sex, or 3) work alone. After 30 minutes, the low-achieving students were given a test on problems similar to those they had worked on in their treatment condition.

Using Analysis Tools for Two-Way Between-Groups ANOVA

1. Enter the data in an Excel worksheet as shown below. It is necessary to utilize labels for the three levels of the treatment variable (Same, Different, Alone) as well as for the two levels of the sex variable (Boy, Girl). In addition, the data need to be sorted by sex. As you can see in the worksheet below, the boys' test scores were entered first and then the girls'. Here, the label **Boy** appears only in the first row of the boys' data and the label **Girl** appears only in the first row of the girls' data. The procedure will also perform accurately if a group label appears in every row of column A.

	A	B	C	D
1	Sex	Same	Different	Alone
2	Boy	7	9	4
3		9	11	5
4		10	14	11
5		12	9	8
6		8	7	2
7	Girl	14	6	10
8		8	4	9
9		11	6	4
10		12	11	9
11		10	3	8

2. Click on **Tools** in the menu Bar and select **Data Analysis**.

If Data Analysis does not appear as a choice in the Tools menu, you will need to load the Microsoft Excel ToolPak add-in. Follow the procedure on page 7.

3. In the Data Analysis dialog box, select **Anova: Two-Factor With Replication** and click **OK**.

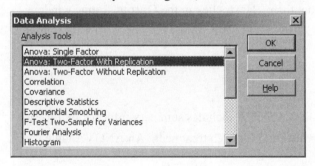

4. Complete the ANOVA dialog box as shown below. Detailed instructions are given immediately after the dialog box.

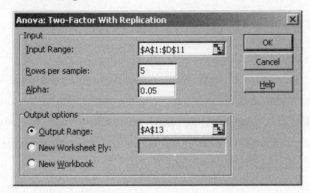

- **Input Range**. Enter the worksheet location of the data. Activate cell **A1** and drag to the end of the data in cell **D11**. You can also key in the data range by entering **A1:D11** or **$A$1:$D$11**.

  Excel will not carry out this analysis procedure unless the range includes labels for the levels of the sex variable and labels for the levels of the treatment variable.

- **Rows per sample**. There must be the same number of rows in each sample. For this research problem, the five rows immediately below the treatment condition labels contain test scores for boys, and the next five rows contain test scores for girls.

  Excel will not perform this analysis if cell n's are unequal or if the data set has any missing values.

- **Alpha**. Let's use the default value of .05.
- **Output options**. Let's place the output in the same worksheet with the data. Click in the button to the left of **Output Range** to place a black dot there. Next, click in the **Output Range** window. Then, click in cell A13 of the worksheet. Cell A13 will be the upper leftmost cell of the output.

5. Click **OK**.

Interpreting the Output

The output provided for the two-factor between-groups ANOVA includes summary information for each gender-treatment combination as well as for each gender and for each treatment. An ANOVA table is also included.

SUMMARY

The SUMMARY portion of the output is displayed at the top of the next page.

		Same	Different	Alone	Total
13	Anova: Two-Factor With Replication				
14					
15	SUMMARY	Same	Different	Alone	Total
16	*Boy*				
17	Count	5	5	5	15
18	Sum	46	50	30	126
19	Average	9.2	10	6	8.4
20	Variance	3.7	7	12.5	9.828571
21					
22	*Girl*				
23	Count	5	5	5	15
24	Sum	55	29	40	124
25	Average	11	5.8	8	8.266667
26	Variance	5	9.7	5.5	10.6381
27					
28	*Total*				
29	Count	10	10	10	
30	Sum	101	79	70	
31	Average	10.1	7.9	7	
32	Variance	4.766667	12.32222	9.111111	

- **BOY**. The top portion of the summary section presents the count, sum, average, and variance for the boys in each of the three treatment conditions. The total column at the far right presents this same information for the boys across all three conditions. For example, the average performance for the boys in the Same, Different, and Alone conditions was 9.2, 10, and 6, respectively. The average score for boys across the three conditions was 8.4.

- **GIRL**. The next portion of the summary section presents the count, sum, average, and variance for the girls in each of the three treatment conditions. The total column at the far right displays this same summary information for the girls across the three conditions. For example, the variance of the girls' scores in the Same, Different, and Alone conditions was 5, 9.7, and 5.5, respectively. The variance of the scores of all 15 girls was 10.6381.

- **Total**. The last part of the summary section presents the count, sum, average, and variance for each treatment condition across sexes. For example, the sum of the scores in the Same, Different, and Alone conditions for the 10 boys and 10 girls combined was 101, 79, and 70, respectively.

ANOVA Table

The ANOVA table portion of the output is displayed below..

		SS	df	MS	F	P-value	F crit
35	ANOVA						
36	*Source of Variation*	SS	df	MS	F	P-value	F crit
37	Sample	0.133333	1	0.133333	0.018433	0.893136	4.259677
38	Columns	50.86667	2	25.43333	3.516129	0.04579	3.402826
39	Interaction	62.06667	2	31.03333	4.290323	0.025528	3.402826
40	Within	173.6	24	7.233333			
41							
42	Total	286.6667	29				

- **Source of Variation**. In the two-way between-groups ANOVA, total variation is partitioned into sample variation (sex), column variation (treatment), interaction variation (sex x treatment interaction), and within groups variation.

- **SS**. Sex is the sample or row factor. I have designated sex as factor A in the sum of squares formulas. Treatment is the column factor and is designated as factor B. The sex x treatment interaction is referred to as A x B. I will first calculate I, II, III, IV, and V, and then use these values to compute each SS.

$$I = \sum_{N=1}^{N} X_N^2 \text{ , where N refers to the total number of observations (N = 30).}$$

$$I = 7^2 + 9^2 + 4^2 + ... + 8^2 = 2370$$

$$II = \frac{\left(\sum_{N=1}^{N} X\right)^2}{N} \text{ , where N refers to the total number of observations.}$$

$$II = \frac{(250)^2}{30} = 2083.3333$$

$$III = \frac{\sum T_{A_i}^2}{n_i} \text{ , where } n_i \text{ is the number of observations in each level of factor A, and } T_{A_i} \text{ is the sum of the}$$

observations in each level of factor A.

$$III = \frac{126^2}{15} + \frac{124^2}{15} = 2083.4666$$

$$IV = \frac{\sum T_{B_j}^2}{n_j} \text{ , where } n_j \text{ is the number of observations in each level of factor B, and } T_{B_j} \text{ is the sum of the}$$

observations in each level of factor B.

$$IV = \frac{101^2}{10} + \frac{79^2}{10} + \frac{70^2}{10} = 2134.2$$

$$V = \frac{\sum T_{A_iB_j}^2}{n_{ij}} \text{ , where } n_{ij} \text{ is the number of observations in each cell, and } T_{A_iB_j} \text{ is the sum of the}$$

observations in each cell.

$$V = \frac{46^2}{5} + \frac{50^2}{5} + \frac{30^2}{5} + ... + \frac{40^2}{5} = 2196.4$$

$$SS_A = III - II = .1333$$

$$SS_B = IV - II = 50.8667$$

$$SS_{AXB} = II + V - III - IV = 62.0667$$

$$SS_{Within} = I - V = 173.6$$

$$SS_{Total} = I - II = 286.6667$$

- **df.** The formulas for df are as follows:

  $df_A = A - 1$, where A is the number of levels of factor A (sex).

  $df_A = 2 - 1 = 1$

  $df_B = B - 1$, where B is the number of levels of factor B (treatment).

  $df_B = 3 - 1 = 2$

  $df_{AXB} = (A - 1)(B - 1)$

  $df_{AXB} = (2 - 1)(3 - 1) = 2$

  $df_{Within} = AB(n_{ij} - 1)$, where n_{ij} is the number of observations in each cell.

  $df_{Within} = (2)(3)(5 - 1) = 24$

  $df_{Total} = N - 1$, where N is the total number of observations.

  $df_{Total} = 30 - 1 = 29$

- **MS**. The mean square terms are found by dividing each SS by its respective df.

- **F**. To obtain the three F-test values, MS_A, MS_B, and MS_{AXB} are each divided by MS_{Within}.

- **P-value**. Compare each P-value with .05 (alpha) to determine whether any of the three effects are statistically significant. For this problem, the F associated with factor B (treatment) and the F associated with the AxB interaction (sex x treatment) are significant because the P-values are both less than .05.

- **F crit**. If any of the three obtained F-test values exceed the critical F-test value, the result is statistically significant. For this research problem, the main effect of B (treatment) and the AxB interaction (sex x treatment) are both significant—the same conclusion that was reached by comparing P-values with alpha.

F-Test for Two Sample Variances

The F-test is most often used in analysis of variance to test the significance of differences among group means. The F-test can also be used to test the hypothesis that two population variances are equal. The formula is

$$F = \frac{S_1^2}{S_2^2}$$

where S_1^2 refers to the variance of sample 1 and S_2^2 refers to the variance of sample 2. Numerator degrees of freedom are equal to $n_1 - 1$, and denominator degrees of freedom are equal to $n_2 - 1$. This F-test is carried out as a one-tailed test, assuming that you have specified in advance which of the two variances is larger. If you want to do a two-tailed test, you would place the larger of the two variances in the numerator and divide the desired alpha level in half. For example, if you wish to do a two-tailed test with alpha equal to .05, set the one-tailed alpha equal to .025.

Assumptions Underlying the F-Test

The assumptions underlying the F-test for two population variances are

1. Samples are randomly selected from the two populations.
2. Observations are normally distributed in the two populations.
3. The data are measured on an interval or ratio scale.

Sample Research Problem

A couple of years ago, I asked the students in my undergraduate statistics classes to provide responses to a survey that asked questions about their "ideal selves," such as ideal weight and ideal number of friends. In one of the questions, I asked the students to estimate the square footage of their ideal home. After looking at some of the responses, I was fairly certain that men were much more accurate in estimating reasonable square footage than women. I decided to carry out an F test to find out if the variances of the men's and women's ideal square footages were equal.

Using Analysis Tools for the F-Test for Two Sample Variances

1. Enter the data into an Excel worksheet as shown below.

	A	B
1	Men	Women
2	3400	40
3	2500	5000
4	1000	400
5	4000	100000
6	900	8000
7	1600	100
8	1800	25000
9	800	3600
10	2000	500

2. Click **Tools** in the Menu Bar and select **Data Analysis**.

If Data Analysis does not appear as a choice in the Tools menu, you will need to load the Microsoft Excel ToolPak add-in. Follow the procedure on page 7.

3. In the Data Analysis dialog box, select **F-Test Two-Sample for Variances** and click **OK**.

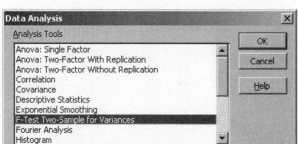

4. Complete the F-Test Two Sample for Variances dialog box as shown below. Detailed information about the entries is provided immediately after the dialog box.

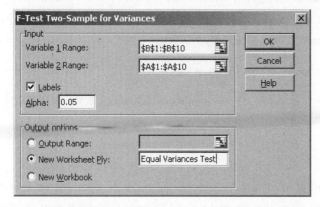

- **Variable 1 Range**. When you are doing a two-tailed test, variable 1 should be the group with the larger variance. So, regardless of whether I am doing a two-tailed or one-tailed test, I usually designate the group with the larger variance as variable 1. For this example, the women's square footages have a larger variance than the men's. So, the women's square footages should be designated as variable 1. Click in the **Variable 1 Range** window. Then click in cell **B1** and drag to cell **B10**. If you prefer, you can manually enter **B1:B10**.

- **Variable 2 Range**. This is the range of the variable that has the smaller of the two group variances— the men's square footage responses. Click in the **Variable 2 Range** window. Then click in cell **A1** and drag to cell **A10**. If you prefer, you can manually enter **A1:A10**.

- **Labels**. The box to the left of **Labels** should be checked if the first row of the input contains labels that should not be included in the data analysis. The first row of our worksheet contains labels (Men and Women), so click in the box to place a check mark there.

- **Alpha**. Alpha refers to the type I error probability for the statistical test. Let's use the default value of .05.

- **Output options**. Let's use the default option of **New Worksheet Ply**. When we select this option, we can give the new worksheet a special name to make it easily identifiable. Let's use the name "Equal Variances Test." Click in the New Worksheet Ply window. Enter **Equal Variances Test**.

5. Click **OK**. The output that you will receive is displayed at the beginning of the next section.

Interpreting the Output

	A	B	C
1	F-Test Two-Sample for Variances		
2			
3		Women	Men
4	Mean	15848.89	2000
5	Variance	1.06E+09	1257500
6	Observations	9	9
7	df	8	8
8	F	841.6199	
9	P(F<=f) one-tail	6.92E-11	
10	F Critical one-tail	3.438101	

- **Mean**. The women's mean ideal square footage response is 15,848.89 and the men's is 2,000.

- **Variance**. The variance of the women's responses is 1.06E+09. Scientific notation is used because the variance is extremely large. If you move the decimal point 9 places to the right (i.e., +09), you see that the variance is 1,060,000,000. The variance of the men's responses is 1,257,500.

- **Observations**. The responses of 9 women and 9 men were included in the analysis.

- **df**. The degrees of freedom (df) for the numerator (women's responses) is equal to $n_1 - 1 = 9 - 1 = 8$. The df for the denominator (men's responses) is equal to $n_2 - 1 = 9 - 1 = 8$.

- **F**. The obtained F is equal to 841.6199.

- **P(F<=f) one-tail**. The chance one-tailed probability of obtained F is 6.92E –11. Scientific notation is used because the probability is extremely small. If you move the decimal point 11 places to the left, you see that the probability is .0000000000692. Because the probability of obtained F is smaller than alpha, the result is statistically significant.

- **F Critical one-tail**. The critical value of F for a one-tailed test is 3.4381. Because obtained F is greater than critical F, the result is statistically significant.

Chapter 10

Correlation

Correlation coefficients are numerical indices that provide information regarding the relationship between two variables. In this chapter, I describe how to use features available in Excel for computing the Pearson correlation coefficient and the Spearman rank correlation. I also provide instructions for generating correlation matrices and creating scatterplots.

Pearson Correlation Coefficient

The Pearson correlation coefficient, symbolized as r, ranges from -1, through 0, to +1. Coefficients close to -1 and +1 indicate strong linear relationships, whereas coefficients close to 0 indicate weak linear relationships. The correlation between two quantitative variables, X and Y, can be found by applying the formula

$$r_{XY} = \frac{Cov(X, Y)}{S_X S_Y}$$

where $Cov(X, Y)$ = the covariance of X and Y,

S_X = the standard deviation of X, and

S_Y = the standard deviation of Y.

Students of statistics may be more familiar with the computational formula, expressed as

$$r_{XY} = \frac{\sum XY - \frac{\left(\sum X\right)\left(\sum Y\right)}{N}}{\sqrt{\left(\sum X^2 - \frac{\left(\sum X\right)^2}{N}\right)\left(\sum Y^2 - \frac{\left(\sum Y\right)^2}{N}\right)}}$$

Sample Research Problem

An investigator was interested in parental influence on daughters regarding achievement in mathematics. Data were collected from 14 randomly selected intact families who had a 16-year-old daughter. Each daughter took a standardized mathematics achievement test. Each parent also completed a mathematics achievement test. The scores were entered in an Excel worksheet as displayed at the top of the next page.

	A	B	C
1	Daughter	Mother	Father
2	84	90	72
3	65	70	85
4	91	86	81
5	75	82	83
6	81	84	84
7	79	93	72
8	83	72	70
9	92	90	88
10	61	74	68
11	73	60	82
12	85	83	72
13	90	91	94
14	54	64	69
15	70	78	62

CORREL Function

Excel's CORREL function returns the value of Pearson's r for any two quantitative variables in a data set. No labels are attached to the r values, however, so I recommend that you enter appropriate labels in the worksheet.

1. Enter the data into an Excel worksheet as shown at the top of this page.

2. Let's find the correlation between daughter's achievement and mother's achievement. Activate cell **A17** and the label **Daughter-Mother**. Press [**Enter**].

Make column A wider so that the entire label will be displayed.

14	54	64	69
15	70	78	62
16			
17	Daughter-Mother		

3. Activate cell **B17** where the r value will be placed.

4. Click **Insert** in the Menu Bar and select **Function**.

5. In the Insert Function dialog box, select the **Statistical** category, select the **CORREL** function, and click **OK**.

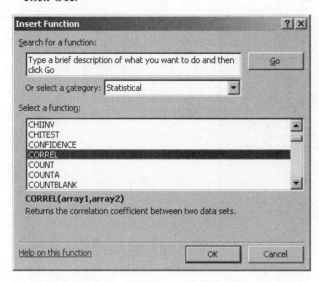

6. The flashing I-beam should be positioned in the **Array 1** window. In the worksheet, click and drag over the range for Daughter, **A1** to **A15**. If you prefer, you can enter the range manually, **A1:A15**.

> *For the CORREL function, inclusion of the column label, cell A1 in this example, is optional.*

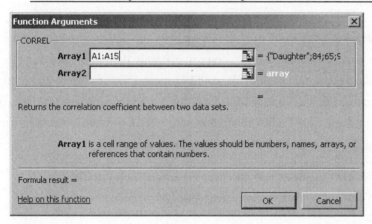

7. Click in the **Array 2** window. In the worksheet, click and drag over the range for Mother, **B1** to **B15**.
    Or, if you prefer, you can manually enter **B1:B15**.

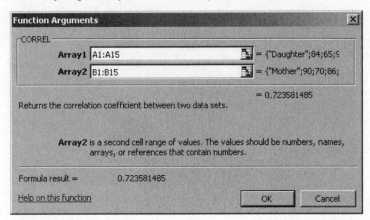

8. Click **OK** and a Pearson correlation coefficient of .7236 is returned and placed in cell B17 of the
    worksheet.

14		54	64	69
15		70	78	62
16				
17	Daughter-Mother	0.723581		

Correlation Matrices

When a data set includes a number of variables, as this one does, it is frequently the desire of the researcher to
calculate not just one Pearson correlation coefficient but several. You can certainly do that by utilizing the
CORREL function; however, it would take a great deal of time. It would be more efficient simply to arrange
the worksheet so that the relevant quantitative variables are in adjacent columns, and then use Data Analysis
Tools to produce a correlation matrix. In this section, I will explain how to use the Correlation Analysis Tool
to compute correlations between all possible pairings of the three variables in the data set.

> *The Correlation Analysis Tool can also be used to obtain the correlation between only one
> pair variables as long as the two variables meet the requirement of being placed in adjacent
> columns in the worksheet.*

1. If you have not already done so, enter the data shown on page 204 in an Excel worksheet.

2. Click on **Tools** in the Menu Bar and select **Data Analysis**.

> *If Data Analysis does not appear as a choice in the Tools menu, you will need to load the
> Microsoft Excel ToolPak add-in. Follow the procedure on page 7.*

3. In the Data Analysis dialog box, select the **Correlation** Analysis Tool and click **OK**.

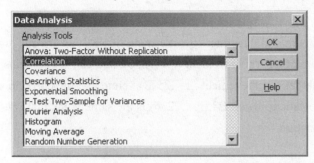

4. Complete the Correlation dialog box as shown below. A description of the entries is given immediately after the dialog box.

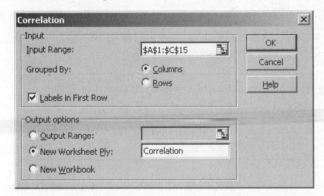

- **Input Range**. Click in the **Input Range** window. Then click on cell **A1** in the worksheet and drag to cell **C15**. If you prefer, you can manually enter **A1:C15**. Note that this range includes the entire data set which has three variables.
- **Grouped by**. The data in the worksheet are grouped by columns. Since **Columns** was already selected, you do not need to do anything additional here.
- **Labels in First Row**. Click in the box next to **Labels in First Row**. The check mark in this box lets Excel know that the first row contains variable labels (e.g., Daughter) that should not be included in the analysis.
- **Output options**. Select **New Worksheet Ply**. This is the default option. Unless specified otherwise, the correlation matrix will be pasted into a new worksheet with A1 as the upper left cell. If you select this option, you may give the worksheet a name that will enable you to identify it easily. Let's name the worksheet **Correlation**.

5. Click **OK**. The correlation matrix is generated and placed in the sheet named Correlation.

	A	B	C	D
1		Daughter	Mother	Father
2	Daughter	1		
3	Mother	0.723581	1	
4	Father	0.466842	0.257273	1

Interpreting the Output

Each entry in the matrix is the correlation between a pair of variables in the data set. For example, cell B2 contains the correlation between daughter's achievement and daughter's achievement (r = 1), cell B3 contains the correlation between daughter's achievement and mother's achievement (r = .7236), and cell C4 contains the correlation between mother's achievement and father's achievement (r= .2573). Note that the correlation between daughter's achievement and mother's achievement in cell B3 (r = .7236) is the same value that was obtained previously when utilizing the CORREL function.

Scatterplot

Excel's Chart Wizard provides an efficient way to produce a scatterplot of two variables. The procedure will not work, however, unless the two variables are adjacent to each other in the worksheet. In this section, I will show you how to use the Chart Wizard to produce a scatterplot of daughter's mathematics achievement and mother's mathematics achievement.

> *If you wish to produce a scatterplot of two variables that are not in adjacent columns of the worksheet, I suggest that you copy the variables of interest into a new worksheet rather than rearranging the entire data set. Cut, paste, or delete as necessary so that the two variables of interest are placed in adjacent columns.*

1. Go back to the worksheet that contains the data. That worksheet is most likely Sheet 1. Click on the **Sheet1** tab at the bottom of the screen.

2. Click **Insert** in the Menu Bar and select **Chart**.

3. **Chart Wizard-Step 1 of 4-Chart Type**. Select **XY(Scatter)** and click **Next>**.

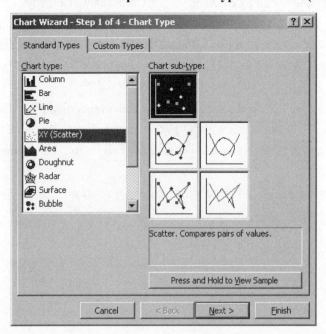

4. **Source Data**. Click and drag over the range of numerical values in columns A and B (**A2** to **B15**). I did not have you include the first row of these columns because you do not need the variable labels for the scatterplot. The range, $A$2:$B$15, appears in the Source Data dialog box. Because the data are in columns, **Columns** should be selected in the **Series in** section at the bottom of the dialog box. Click **Next>**.

> *My data are located in Sheet 1, so you see Sheet 1 in the Data range window in the dialog box shown below. If your data are in a different worksheet, a different sheet number will be displayed.*

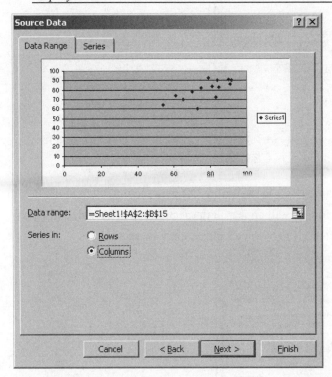

5. **Chart Wizard-Step 3 of 4-Chart Options**. If the Titles page is not displayed, click the **Titles** tab at the top of the dialog box. Type the headings in the appropriate windows of the dialog box as shown at the top of the next page.

Chart title: Mathematics Achievement of Daughters and Their Mothers

Value (X) axis. Daughter

Value (Y) axis. Mother

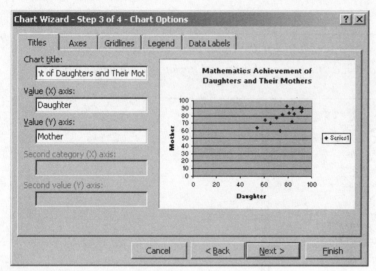

6. Click the **Gridlines** tab at the top of the dialog box. In the **Value (X) axis** section, click in the box to the left of **Major gridlines** so that these gridlines will appear in the scatterplot. The Value (Y) axis major gridlines should have been pre-selected. If not, place a check mark there also.

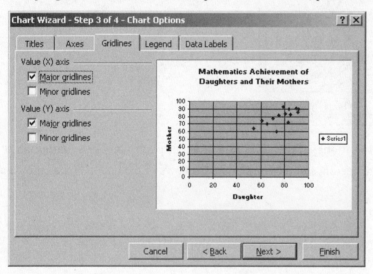

7. Click the **Legend** tab at the top of the dialog box. You want to remove the Series 1 legend that is displayed at the right side of the graph. Click in the box to the left of **Show legend** to remove the check mark that appears there. Click **Next>**.

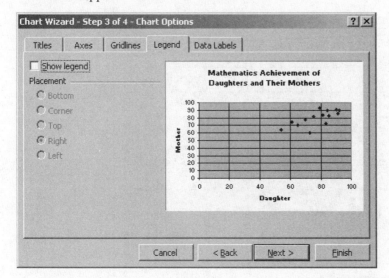

8. **Chart Wizard-Step 4 of 4-Chart Location**. For this example, we would like a large scatterplot that will be displayed on a full sheet of paper. So, select **As new sheet**. Click **Finish**.

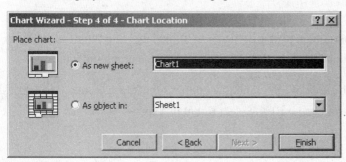

The completed scatterplot is shown below.

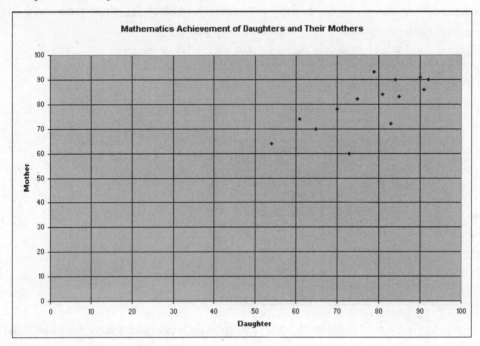

Modifying the Scatterplot

Although this scatterplot is accurate, it is unattractive, primarily because of the large amount of blank space to the left of the scatter of points. You can modify this as well as other aspects of the graph quite easily.

1. **Format X Axis**. Move the pointer to the area of the graph that contains the X-axis scale (0 to 100). Right-click on any x-axis value. I right-clicked on 10. Select **Format Axis**.

2. Click on the **Scale** tab. Click in the box to the left of **Minimum** to "unselect" Auto. The lowest achievement test score for the daughters in the data set is 54, so let's use 50 as the minimum on the X-axis scale. Enter **50** in the space to the right of Minimum. Click **OK**.

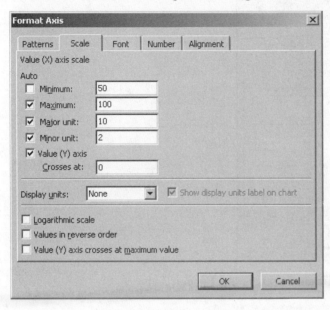

The scatterplot with the X-axis modification is shown below.

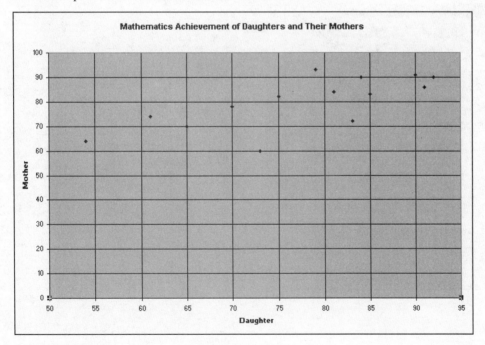

3. **Font**. Let's select a larger size font for the Y-axis and X-axis scales. Right-click on any X-axis value. I right-clicked on 50. Select **Format Axis**.

4. Click on the **Font** tab at the top of the Format Axis dialog box.

5.   Select font size **12**. Click **OK**.

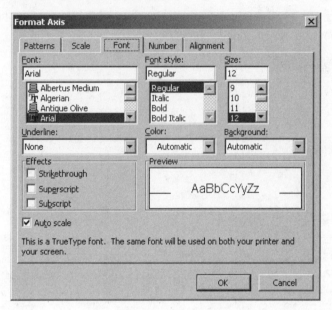

6.   Right-click on any Y-axis value. I right-clicked on 40. Select **Format Axis**.

7.   Select font size **12**. Click **OK**.

8.   The dots are rather hard to see. Let's make them bigger. Right-click on any dot. Select **Format Data Series**.

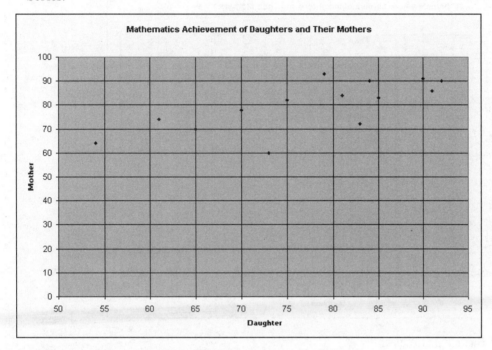

9.   Click the up arrow to change the dot size to **10**. Click **OK**.

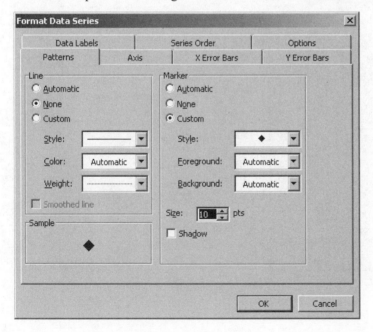

10.   **Trendline**. Let's add a trendline to the graph. Right-click on any dot. Select **Add Trendline**.

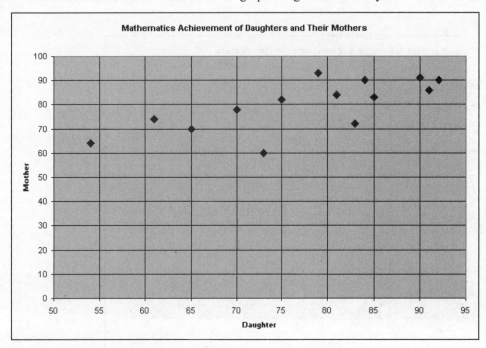

11.   Click on the **Linear** type to select it.

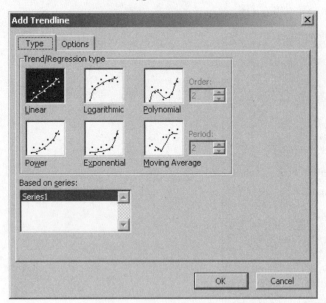

12. **Display R².** Let's display the R^2 value on the graph. Click the **Options** tab at the top of the Format Axis dialog box.

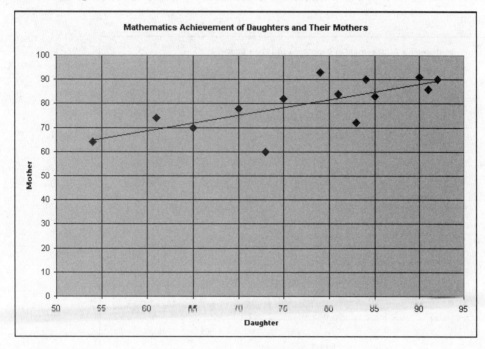

13. Click in the box to the left of **Display R-squared value on chart** so that a check mark appears there. Click **OK**.

14. My R^2 value (.5236) was placed in the upper right hand corner and was difficult to read. I decided to move it to a different location. If you also would like to move the R^2 value, just click on it and drag it to a better location.

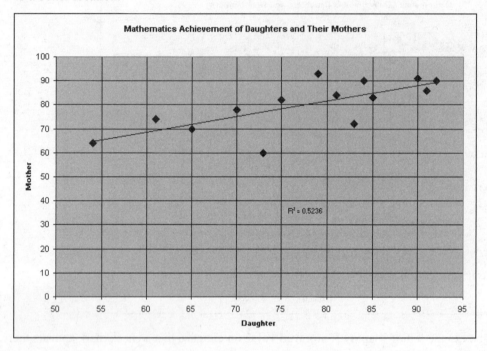

15. **Font size**. Let's increase the font size for all the titles so that they are easier to read. Start with the title at the top. Click directly on **Mathematics Achievement of Daughters and Their Mothers** and a grey border will appear. Click **Format** in the Menu Bar and select **Selected Chart Title**.

16. Click the **Font** tab at the top of the Format Chart Title dialog box. Change the Font style to **Bold Italic**. Change the Size to **16**. Click **OK**.

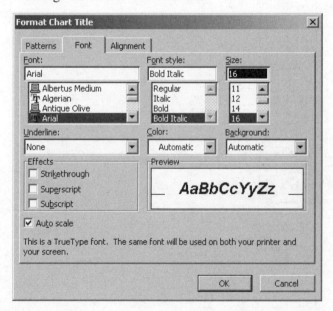

17. Click directly on **Mother**. Click **Format** in the Menu Bar. Select **Selected Axis Title**.

18. Change the Size to **12**. Click **OK**.

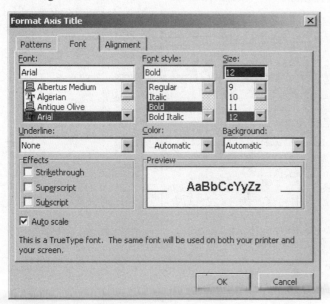

19. Click directly on **Daughter**. Click **Format** in the Menu bar. Select **Selected Axis Title**.

20. Change the size to **12**. Click **OK**.

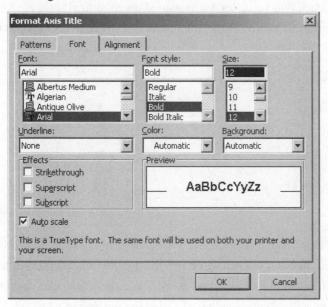

The completed scatterplot is displayed below.

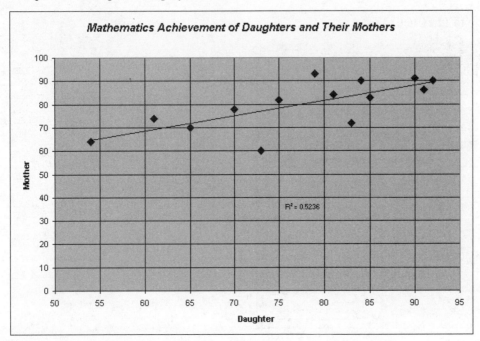

Spearman Rank Correlation Coefficient

Spearman's rank correlation coefficient is used when two variables are measured on an ordinal scale or two variables have been converted to ranks. The coefficient expresses the amount of agreement in the two sets of measures.

The formula for computing Spearman's correlation is

$$r_s = \frac{6 \sum D^2}{N(N^2 - 1)}$$

where   D = the difference in the ranks, and

N = the number of pairs of ranks.

Sample Research Problem

A researcher was interested in the extent to which men and women agreed on the quality of beers produced by microbreweries. The researcher prepared a list of seven microbrews and asked a sample of men and women to vote for the one that they considered the best overall. The researcher recorded the number of votes received by each beer.

RANK Function and Correlation Analysis Tool

1.   Begin by entering the data in an Excel worksheet as shown at the top of the next page.

	A	B	C
1	Beer	Women	Men
2	Big Dog	119	104
3	Honey Lager	92	72
4	Blue Frog	83	94
5	Rocky Brew	61	24
6	Sherwood Forest	31	10
7	Stone Mountain	24	18
8	Maple Brau	5	14

2. The data need to be converted to ranks. Key in column labels of **W Rank** and **M Rank** in cells D1 and E1, respectively, for the women's and men's ranks.

	A	B	C	D	E
1	Beer	Women	Men	W Rank	M Rank
2	Big Dog	119	104		
3	Honey Lager	92	72		

3. First, let's generate the ranks for the women's data. Click in cell **D2** to activate that cell. Then click on **Insert** in the Menu Bar and select **Function**.

4. In the Insert Function dialog box, select **Statistical** in the category window and select the **RANK** function. Click **OK**.

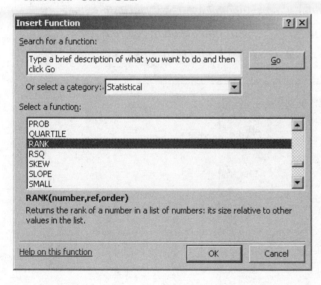

5. Click in the **Number** window and enter **B2**. Cell B2 is the worksheet address of the number of votes that Big Dog received from women.

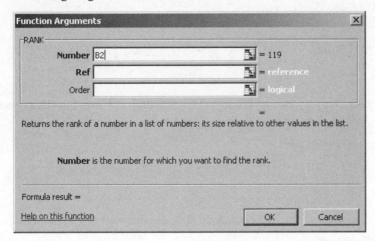

6. Click in the **Ref** window and enter **$B$2:$B$8**. What you are telling Excel to do is to find the rank of the number in cell B2 based on its position within the referenced range. Click **OK**.

> *It is very important that you include the dollar signs in the Ref window. The dollar signs make the range of cells absolute rather than relative. You want absolute cell references, because, in the next step, you will be copying the function in cell D2 to cells D3 through D8, and you want to use the same reference for all observations.*

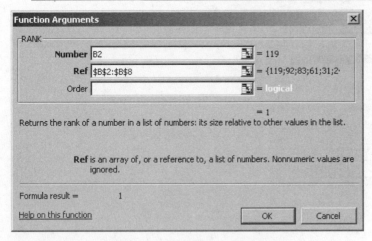

7. Copy the contents of cell D2 to cells D3 through D8.

	A	B	C	D	E
1	Beer	Women	Men	W Rank	M Rank
2	Big Dog	119	104	1	
3	Honey Lager	92	72	2	
4	Blue Frog	83	94	3	
5	Rocky Brew	61	24	4	
6	Sherwood Forest	31	10	5	
7	Stone Mountain	24	18	6	
8	Maple Brau	5	14	7	

8.   Click in cell **E2**. Then click **Insert** in the Menu Bar and select **Function**.

9.   In the Insert Function dialog box, select **Statistical** and **RANK**. Click **OK**.

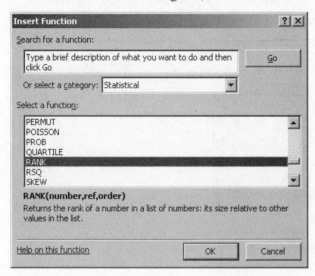

10.   Click in the **Number** window of the RANK dialog box and enter **C2**.

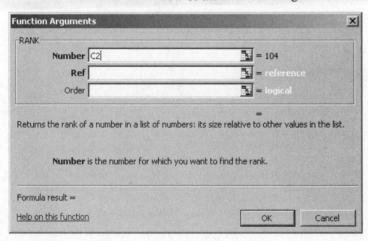

11. Click in the Ref window and enter **$C$2:$C$8**. Click **OK**.

---

*The F4 key will automatically insert dollar signs in a highlighted range. For this example, first enter C2:C8 in the Ref window by clicking and dragging over the range in the worksheet. Next, in the Ref window, select C2:C8 so that it is highlighted. Then press the **F4** key. You should now see $C$2:$C$8.*

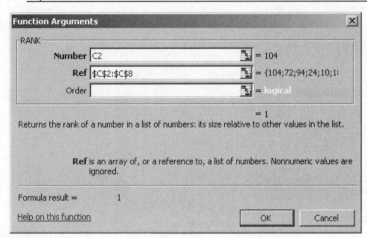

12. Copy the contents of cell E2 to cells E3 through E8. Your completed outlook should look like the output displayed below.

	A	B	C	D	E
1	Beer	Women	Men	W Rank	M Rank
2	Big Dog	119	104	1	1
3	Honey Lager	92	72	2	3
4	Blue Frog	83	94	3	2
5	Rocky Brew	61	24	4	4
6	Sherwood Forest	31	10	5	7
7	Stone Mountain	24	18	6	5
8	Maple Brau	5	14	7	6

13. Rather than using the formula for Spearman's correlation that was presented earlier, you can use the Data Analysis Tool named Correlation to obtain the Pearson correlation coefficient. The results will be identical to those obtained by using Spearman's formula. Click on **Tools** and select **Data Analysis**.

If Data Analysis does not appear as a choice in the Tools menu, you will need to load the Microsoft Excel ToolPak add-in. Follow the procedure on page 7.

---

14. In the Analysis Tools window, select **Correlation** and click **OK**.

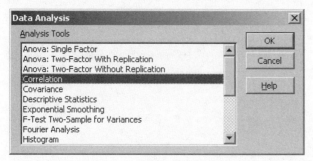

15. Complete the Correlation dialog box as shown below. Detailed information about the entries is provided immediately after the dialog box.

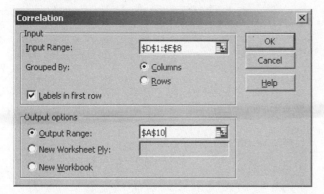

- **Input Range**. Enter the location of the ranks. Click in the **Input Range** window. Then activate cell **D1** and drag to the end of the rank data in cell **E8**. If you prefer, you can manually enter **D1:E8**.
- **Grouped by**. The data are grouped by columns in the worksheet. **Columns** is the default setting.
- **Labels in First Row**. A check mark must appear in this box to let Excel know that the top cell in each column of the input range contains a label and not a value that is to be included in the analysis.
- **Output options**. Select **Output Range** to place the output in the same worksheet as the data. Click in the Output Range window. Then enter cell **A10** in the window. A10 will be the uppermost left cell of the correlation output.

16. Click **OK**. The output you will receive is shown below.

	A	B	C	D	E
1	Beer	Women	Men	W Rank	M Rank
2	Big Dog	119	104	1	1
3	Honey Lager	92	72	2	3
4	Blue Frog	83	94	3	2
5	Rocky Brew	61	24	4	4
6	Sherwood Forest	31	10	5	7
7	Stone Mountain	24	18	6	5
8	Maple Brau	5	14	7	6
9					
10		W Rank	M Rank		
11	W Rank	1			
12	M Rank	0.857143	1		

Interpreting the Output

The Spearman rank correlation coefficient is .8571. This indicates fairly strong agreement between men and women regarding the quality of microbrews.

Chapter 11

Regression

Two-Variable Regression

In a two-variable regression investigation, we have one predictor variable (X) that is used to predict a second variable (Y). The variable that is being predicted (Y) is referred to as the criterion variable. The prediction equation (also known as the regression line equation) is given by

$$\hat{Y} = a + bX$$

where $\hat{Y}$ = the predicted value of Y,

   a = the Y-axis intercept,

   b = the slope of the regression line, and

   X = a value of the predictor variable.

The formula for computation of the intercept is

$$a = \overline{Y} - b\overline{X}$$

where $\overline{Y}$ = the mean of Y, and

   $\overline{X}$ = the mean of X.

The formula for computation of the slope is

$$b = r_{XY} \cdot \frac{S_Y}{S_X}$$

where r_{XY} = the correlation between X and Y,

   S_Y = the standard deviation of Y, and

   S_X = the standard deviation of X.

Sample Research Problem

A university was interested in finding out how well ACT scores predicted the first-semester GPA of its freshmen. The worksheet containing the data for 15 freshmen is shown at the top of the next page.

	A	B	C
1	Subject	ACT	GPA
2	1	24	3.25
3	2	21	2.87
4	3	18	2.66
5	4	22	3.33
6	5	22	2.87
7	6	22	3.21
8	7	18	2.76
9	8	28	3.91
10	9	29	3.55
11	10	18	2.55
12	11	20	2.44
13	12	24	3.22
14	13	25	3.22
15	14	24	3.44
16	15	21	3.01

Steps to Follow to Analyze the Sample Research Problem

1. Enter the data into an Excel worksheet as shown above.

2. In the Menu Bar, click on **Tools** and select **Data Analysis**.

> *If Data Analysis does not appear as a choice in the Tools menu, you will need to load the Microsoft Excel ToolPak add-in. Follow the procedure on page 7.*

3. In the Data Analysis dialog box, select **Regression** and click **OK**.

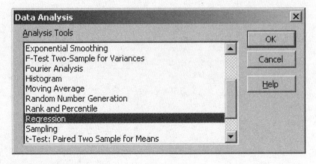

4. Complete the Regression dialog box as shown below. A description of the entries follows immediately after the dialog box.

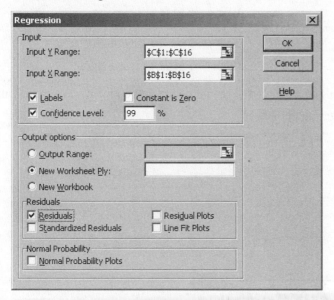

- **Input Y Range** is the range in the worksheet containing scores on the criterion variable (GPA). To enter the range, first click in the **Input Y Range** window and then drag over cells **C1** through **C16**. If you prefer, the range can be entered manually by keying in **C1:C16**.

- **Input X Range** is the range for scores on the predictor variable (ACT). First click in the **Input X Range** window and then drag over the cells **B1** through **B16**. If you prefer, you can manually enter **B1:B16**.

- **Labels** should be checked because the variable names in cells B1 (ACT) and C1 (GPA) have been included in the input ranges and should not be included in the regression analysis. These labels, however, will be used in the output.

- **Constant is Zero** is not checked, because you do not want to force the regression line through the origin.

- **Confidence Level** is checked. Enter a value of **99** in the space to the right of Confidence Level. You will now see both 95% and 99% boundaries reported in the regression output. If Confidence Level is not checked, then only the default value of 95 will be utilized, and you will see the 95% boundaries reported twice in the output.

- **Output options** provide three choices. When you select an option, take into consideration that the regression output will take up at least seven columns.
    - **Output Range** is used when you want the regression output placed in the same worksheet as the data.
    - **New Worksheet Ply** will place the regression output in a new worksheet within the same book of sheets. If you are currently on sheet 1, this option will allow you to place the data on another sheet, with A1 as the upper left cell of the output.

• **New Workbook** will place the regression output in a new workbook, with A1 as the upper left cell.

Click on Help in the Regression dialog box to obtain a description of the options and required entries.

5. Click **OK** and the output shown below will be generated. I recommend that you increase column width as necessary so that you can read the longer output labels.

	A	B	C	D	E	F	G	H	I
1	SUMMARY OUTPUT								
2									
3	*Regression Statistics*								
4	Multiple R	0.8835824							
5	R Square	0.7807178							
6	Adjusted R Square	0.7638499							
7	Standard Error	0.1947608							
8	Observations	15							
9									
10	ANOVA								
11		*df*	*SS*	*MS*	*F*	*Significance F*			
12	Regression	1	1.755646929	1.755647	46.28433	1.256E-05			
13	Residual	13	0.493113071	0.037932					
14	Total	14	2.24876						
15									
16		*Coefficients*	*Standard Error*	*t Stat*	*P-value*	*Lower 95%*	*Upper 95%*	*Lower 99.0%*	*Upper 99.0%*
17	Intercept	0.7217766	0.3511329	2.055565	0.060486	-0.0367999	1.4803532	-0.3359325	1.7794858
18	ACT	0.1055457	0.015513989	6.803259	1.26E-05	0.0720298	0.1390616	0.05881327	0.1522781

	A	B	C
22	RESIDUAL OUTPUT		
23			
24	*Observation*	*Predicted GPA*	*Residuals*
25	1	3.254873096	-0.004873096
26	2	2.938236041	-0.068236041
27	3	2.621598985	0.038401015
28	4	3.043781726	0.286218274
29	5	3.043781726	-0.173781726
30	6	3.043781726	0.166218274
31	7	2.621598985	0.138401015
32	8	3.677055838	0.232944162
33	9	3.782601523	-0.232601523
34	10	2.621598985	-0.071598985
35	11	2.832690355	-0.392690355
36	12	3.254873096	-0.034873096
37	13	3.360418782	-0.140418782
38	14	3.254873096	0.185126904
39	15	2.938236041	0.071763959

Interpreting the Output

SUMMARY OUTPUT

Regression Statistics

• **Multiple R**. Multiple R is the correlation between the predictor variable(s) and the criterion variable. Because we have only one predictor variable (ACT), .8836 is the Pearson correlation coefficient, r, expressing the linear relationship between ACT and GPA.

- **R Square**. R Square is also referred to as the coefficient of determination. It represents the proportion of variation in Y that is explained by its linear relationship with X. For our two-variable problem, R^2 is equal to .7807. Because we have only one predictor variable, R^2 could be represented as r^2.

- **Adjusted R Square**. The sample R^2 tends to be an optimistic estimate of the fit between the model and the population. Adjusted R^2 generally provides a better estimate. The adjusted r^2 for our sample two-variable problem is .7638.

- **Standard Error** is the standard error of estimate and is interpreted as the average error in predicting Y by means of the regression equation. For the sample problem, we would estimate that we err, on average, .1948 grade points when we use the regression equation to predict GPA from ACT.

- **Observations**. The number of observations refers to the number of subjects included in the analysis. Our analysis was carried out on the data provided by 15 subjects.

ANOVA

Regression analysis includes a test of the hypothesis that the slope of the regression line is equal to 0. If the slope is significantly different from 0, then we conclude that there is a statistically significant linear relationship between ACT and GPA.

- **Regression.** This component represents the variation in GPA that is explained by its relationship with ACT.

- **Residual.** Residual variation represents the variation in GPA that is not explained by ACT. It is considered "error variation" because it is unexplained by the predictor variable we have included in the analysis.

- **Total.** Total refers to "total variation." For this analysis, total variation is partitioned into regression variation and residual variation. Total variation, therefore, is the sum of regression variation and residual variation.

For each source of variation, the output provides degrees of freedom (df) and sums of squares (SS). The F value is obtained by dividing mean square (MS) regression by MS residual. The significance of F is the probability (P-value) associated with the obtained value of F. The test is statistically significant with alpha equal to .05 or .01, because the P-value of 1.25624E-05 is less than either of these values.

Coefficients

The information provided at the bottom of the output refers to the coefficients in the regression equation. The regression equation for the sample research problem is given by

$$\hat{Y} = .7218 + .1055X$$

If we utilize this formula to predict the freshman GPA for a student whose ACT score is 20, we would obtain

$$\hat{Y} = .7218 + .1055(20) = 2.83$$

- **Intercept.** The intercept is .7218. The **t Stat** refers to a test of the hypothesis that the intercept is significantly different from zero. The **P-value** is the probability associated with the obtained t statistic. The 95% and 99% confidence interval boundaries would be applied, respectively, to form a 95% and a 99% confidence interval around the intercept.

- **ACT.** The slope of the regression line is .1055. The **t Stat** refers to a test of the hypothesis that the slope is significantly different from zero. For the two-variable analysis, this t statistic and the F in the ANOVA table provide a test of the same hypothesis. Accordingly, F can be found by squaring the obtained t statistic. The **P-value** is the probability associated with the obtained t statistic. The 95% and 99% confidence interval boundaries can be used to form 95% and 99% confidence intervals, respectively, around the slope.

RESIDUAL OUTPUT

- **Observation**. The observations are numbered from 1 to 15. These numbers correspond to the order in which the observations appear in the data set. For example, observation 1 refers to subject 1 who received an ACT of 24 and a GPA of 3.25.

- **Predicted GPA**. The predicted GPA for each subject using the prediction equation $\hat{Y} = .7218 + .1055X$.

- **Residuals**. This is prediction error and is calculated $Y - \hat{Y}$. For example, the actual GPA, or Y, for observation 1 was 3.25. The predicted GPA, or $\hat{Y}$, for observation 1 was 3.2549. Prediction error for observation 1 was $3.25 - 3.2549$, or -0.0049.

Multiple Regression

Multiple regression extends the concepts of two-variable regression to situations that include one criterion variable (Y) and two or more predictor variables, $X_1, X_2, X_3, \ldots, X_K$. Excel will allow you to include as many as 16 predictor variables. A model that includes two predictor variables is given by

$$\hat{Y} = B_0 + B_1 X_1 + B_2 X_2$$

Sample Research Problem

A personnel researcher wanted to find out how well sales volume of salespersons could be predicted from a score derived from demographic information (Biodata Score) and a Sales Aptitude test score. If the prediction equation worked well for individuals currently employed as salespersons, the equation could presumably be used for the selection of new employees. The data for 14 salespersons were entered in an Excel worksheet.

The predictor variables (Biodata Score and Sales Aptitude) must be entered in adjacent columns of the worksheet.

	A	B	C
1	Biodata Score	Sales Aptitude	Sales (000s)
2	25	35	234
3	33	44	124
4	44	65	345
5	54	55	467
6	45	60	432
7	23	33	235
8	53	54	555
9	22	12	146
10	42	34	345
11	25	23	366
12	16	22	156
13	23	16	123
14	36	55	564
15	33	28	237

Steps to Follow to Analyze the Sample Research Problem

1. Enter the data in an Excel worksheet as shown above.

2. Click on **Tools** and select **Data Analysis**.

   If Data Analysis does not appear as a choice in the Tools menu, you will need to load the Microsoft Excel ToolPak add-in. Follow the procedure on page 7.

3. In the Data Analysis dialog box, select **Regression** and click **OK**.

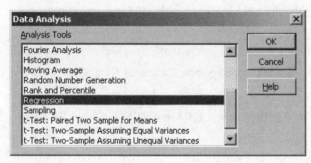

4. A completed Regression dialog box is shown below. A description of the entries follows immediately
    after the dialog box.

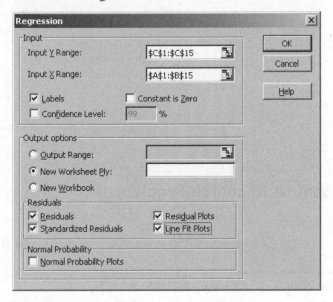

- **Input Y Range** is the range of the criterion variable, Sales(000s). First click in the **Input Y Range**
  window. Then click in cell **C1** in the worksheet and drag to cell **C15**. If you prefer, you can
  manually enter **C1:C15**. Note that the variable label in C1 was included in the input range.
- **Input X Range** includes scores on both predictor variables—Biodata Score and Sales Aptitude.
  Click in the **Input X Range** window. Then click in cell **A1** in the worksheet and drag to cell **B15**. If
  you prefer, you can manually enter **A1:B15**. Note that the variables labels in A1 and B1 were
  included in the input range.
- **Labels** is checked because the variable labels in cells A1, B1, and C1 were included in the data
  ranges.
- **Constant is Zero** is not checked. You do not want to force the regression line through the origin.
- **Confidence Level** is not checked. The default of 95% will be displayed twice in the output.
- Click in the button to the left of **New Worksheet Ply** to place the output in the same workbook but on
  a different sheet.
- **Residuals** are the differences between the actual data points and the values predicted by the
  regression equation. Click in the boxes to select all the residual output options.

---

*Click on Help in the Regression dialog box to obtain a description of the options and
required entries.*

---

5. Click **OK**. The Summary Output provided by the regression procedure includes four sections:
    regression statistics, ANOVA summary table, coefficient summary information, and summary
    information regarding residuals. The lower 95% and upper 95% coefficient boundaries are repeated in
    the right-most sections of the output and are not included here.

	A	B	C	D	E	F	G
1	SUMMARY OUTPUT						
2							
3	*Regression Statistics*						
4	Multiple R	0.7743513					
5	R Square	0.5996199					
6	Adjusted R Square	0.5268235					
7	Standard Error	105.63224					
8	Observations	14					
9							
10	ANOVA						
11		*df*	*SS*	*MS*	*F*	*Significance F*	
12	Regression	2	183818.4963	91909.25	8.236947	0.006510263	
13	Residual	11	122739.8608	11158.17			
14	Total	13	306558.3571				
15							
16		*Coefficients*	*Standard Error*	*t Stat*	*P-value*	*Lower 95%*	*Upper 95%*
17	Intercept	-11.32534	86.32271854	-0.1312	0.897988	-201.3203625	178.669682
18	Biodata Score	6.4876211	4.145268117	1.565067	0.145863	-2.636052528	15.6112947
19	Sales Aptitude	2.6351164	2.921508622	0.901971	0.386396	-3.795080754	9.06531348

Interpreting the Output

Regression Statistics

- **Multiple R** The multiple correlation, R, between the two X variables (Biodata Score and Sales Aptitude) and Y [Sales(000s)] is .7744.

- **R Square**. R^2 is equal to .5996. Therefore, nearly 60 percent of the variation in Y is explainable by its linear relation with the two predictors.

- **Adjusted R Square**. As in the bivariate case, this is an adjusted value that more closely reflects the degree of fit in the population. For our sample problem, adjusted R^2 is equal to .5268.

- **Standard Error**. The standard error of estimate is equal to 105.6322. This is the average amount by which we err if we use the regression equation to predict Sales(000s).

- **Observations**. The data set includes 14 subjects with three variables recorded for each subject.

ANOVA

The ANOVA output is for a statistical test of the hypothesis of no linear relationship between the predictor variables and the criterion variable.

- **Regression**. The variation in Y that is explained by its relationship with the two predictor variables is referred to as *regression*.

- **Residual.** The variation in Y that is not explained by the two predictor variables is referred to as *residual*.

- **Total.** Total variation is partitioned into regression variation and residual variation.

MS regression and MS residual are found by dividing the appropriate SS by the respective df. The F value is obtained by dividing MS regression by MS residual. The significance of F refers to the probability (P-value) associated with obtained F. Statistical significance can be determined by comparing alpha with the P-value of

.0065. The test would be considered significant with alpha equal to .05, as well as .01, because .0065 is less than either of these values.

Coefficient Summary Information

- **Intercept.** The intercept of the regression equation is –11.3253.

- **Biodata Score beta weight.** The beta weight for the Biodata Score is 6.4876.

- **Sales Aptitude beta weight.** The beta weight for Sales Aptitude is 2.6351.

When these values are placed in the regression equation, we see that the equation for predicting Sales(000s) from the two predictor variables is given by

$$\hat{Y} = -11.3253 + 6.4876X_1 + 2.6351X_2$$

If we use the data for the first person in the data set, we obtain

Predicted Sales (in 000s) = –11.3253 + 6.4876(25) + 2.6351(35) = 243.0942.

- The **Standard Error** values, **t Stat**, **P-value**, and **Lower and Upper 95%** confidence interval boundaries are also provided for each of these coefficients. The t Stat for the intercept provides a statistical test of the hypothesis that the intercept is equal to zero. The t Stats for the predictor variable beta weights both test the hypothesis that the beta weight is equal to zero.

Residuals

The residual output is shown below.

	A	B	C	D
23	RESIDUAL OUTPUT			
24				
25	Observation	Predicted Sales (000s)	Residuals	Standard Residuals
26	1	243.0942596	-9.094259552	-0.093593612
27	2	318.7112754	-194.7112754	-2.003871943
28	3	445.4125509	-100.4125509	-1.033396207
29	4	483.9375981	-16.93759806	-0.174313365
30	5	438.7245902	-6.724590194	-0.069206149
31	6	224.8487847	10.15121533	0.104471277
32	7	474.8148606	80.18513938	0.825225713
33	8	163.0237199	-17.02371994	-0.175199688
34	9	350.7487015	-5.748701492	-0.059162787
35	10	211.4728632	154.5271368	1.590316704
36	11	150.4491571	5.550842874	0.057126524
37	12	180.0518065	-57.05180648	-0.587148916
38	13	367.1604187	196.8395813	2.025775412
39	14	276.5494136	-39.54941362	-0.407022963

- **Predicted Sales(000s).** Predicted Sales is the value we will obtain when we utilize the prediction equation. Predicted Sales for the first person in the data set is 243.0943.

- **Residuals.** Residuals are found by subtracting the predicted value from the actual value. For the first person, 234 – 243.0943 = –9.0943.

- **Standard Residuals.** Standard Residuals are found by transforming the residuals to a unit normal distribution with a mean of 0 and standard deviation of 1.

- **Residual Plots.** Residual plots were requested in the output options section of the Regression dialog box. These are shown below. These plots graphically compare actual Y with predicted Y for each of the two predictors. The charts can be selected, reshaped and printed on a full page if desired.

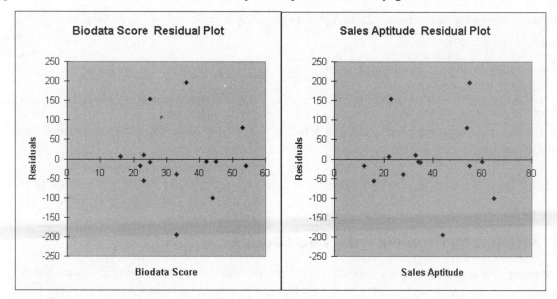

- **Line Fit Plots.** The optional line fit plots were also requested. When we have two predictors, two line fit plots are created. Each plot takes one view of the plane that is generated by the regression equation. The difference between the actual and predicted values of Y is shown for the Biodata Scores in one of the plots and for Sales Aptitude Scores in the other plot. The line fit plots are displayed below.

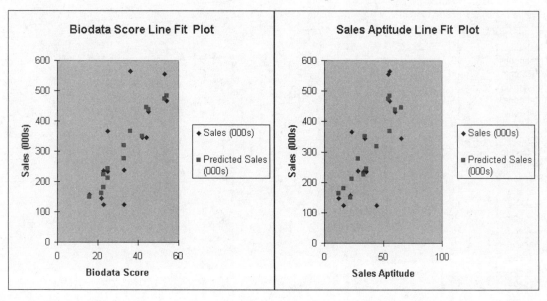

Dummy Coding of Qualitative Variables

When regression analysis is used for analysis of variance, the qualitative group membership variable is introduced into the regression equation by quantifying it. To quantify the group membership variable, first count the number of levels. If the number of levels is equal to K, then the degrees of freedom associated with the group membership variable is K-1. Therefore, K-1 linearly independent variables $(X_1, X_2, X_3, ..., X_{K-1})$ can be introduced into the regression model to represent the group membership variable. The general procedure is:

X_1 = 1 if the observation is a member of level 1, and X_1 = 0 if the observation is not a member of level l.

X_2 = 1 if the observation is a member of level 2, and X_2 = 0 if the observation is not a member of level 2.

X_3 = 1 if the observation is a member of level 3, and X_3 = 0 if the observation is not a member of level 3, and so on up to X_{K-1}.

X_{K-1} = 1 if the observation is a member of level K – 1, and X_{K-1} = 0 if the observation is not a member of level K – 1.

Sample Research Problem with Two Groups

For a two-group problem, I will use the same data that were used for the t-test for two independent samples on page 170. This will not only illustrate dummy coding for a variable with two levels but will also demonstrate that the regression analysis outcome is identical to that of the independent samples t-test.

Steps to Follow to Analyze the Sample Research Problem

1. Enter the data in an Excel worksheet as shown below.

	A	B
1	Treatment	Dreams
2	Melatonin	21
3	Melatonin	18
4	Melatonin	14
5	Melatonin	20
6	Melatonin	11
7	Melatonin	19
8	Melatonin	8
9	Melatonin	12
10	Melatonin	13
11	Melatonin	15
12	Placebo	12
13	Placebo	14
14	Placebo	10
15	Placebo	8
16	Placebo	16
17	Placebo	5
18	Placebo	3
19	Placebo	9
20	Placebo	11

2. Because treatment has two levels, only one dummy variable can be defined. Let's call that variable X1 and place it in column C.

The X1 values can be keyed in directly or you can generate them by using an IF statement. In cell C2, enter =IF(A2="Melatonin",1,0). Press [Enter]. Then copy the contents of cell C2 to cells C3 through C20.

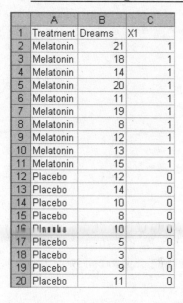

	A	B	C
1	Treatment	Dreams	X1
2	Melatonin	21	1
3	Melatonin	18	1
4	Melatonin	14	1
5	Melatonin	20	1
6	Melatonin	11	1
7	Melatonin	19	1
8	Melatonin	8	1
9	Melatonin	12	1
10	Melatonin	13	1
11	Melatonin	15	1
12	Placebo	12	0
13	Placebo	14	0
14	Placebo	10	0
15	Placebo	8	0
16	Placebo	10	0
17	Placebo	5	0
18	Placebo	3	0
19	Placebo	9	0
20	Placebo	11	0

3. In the Menu Bar, click on **Tools** and select **Data Analysis**.

If Data Analysis does not appear as a choice in the Tools menu, you will need to load the Microsoft Excel ToolPak add-in. Follow the procedure on page 7.

---

4. In the Analysis Tools window, select **Regression** and click **OK**

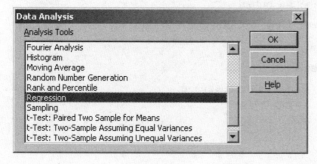

5. Complete the Regression dialog box as shown below. Detailed instructions are given immediately following the dialog box.

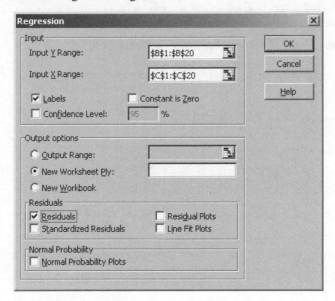

- **Input Y Range**. Enter the range of the Dreams variable in the Input Y Range window. First click in the **Input Y Range** window. Then click in cell **B1** in the worksheet and drag to cell **B20**. If you prefer, you can manually enter **B1:B20**.
- **Input X Range**. Enter the range of the X1 variable in the Input X Range window. First click in the **Input X Range** window. Then click in cell **C1** in the worksheet and drag to cell **C20**. If you prefer, you can manually enter **C1:C20**.
- **Labels**. Click in the box next to **Labels** to place a check mark there. Cells B1 and C1 contain the labels "Dreams" and "X1," respectively. These labels should not be included in the regression analysis. They will, however, be used to label the output.
- **New Worksheet Ply**. Click the button next to **New Worksheet Ply** to place the output on a sheet different from the one that contains the data.
- **Residuals**. Click in the box next to **Residuals**.

6. Click **OK**. The output that you will obtain is shown on the next page.

	A	B	C	D	E	F	G
1	SUMMARY OUTPUT						
2							
3	*Regression Statistics*						
4	Multiple R	0.555618656					
5	R Square	0.308712091					
6	Adjusted R Square	0.268048096					
7	Standard Error	4.204028569					
8	Observations	19					
9							
10	ANOVA						
11		*df*	*SS*	*MS*	*F*	*Significance F*	
12	Regression	1	134.1760234	134.176	7.59178	0.013516432	
13	Residual	17	300.4555556	17.6739			
14	Total	18	434.6315789				
15							
16		*Coefficients*	*Standard Error*	*t Stat*	*P-value*	*Lower 95%*	*Upper 95%*
17	Intercept	9.777777778	1.401342856	6.97743	2.23E-06	6.821202817	12.734353
18	X1	5.322222222	1.931617825	2.75532	0.013516	1.246864882	9.3975796
19							
20							

	A	B	C
21			
22	RESIDUAL OUTPUT		
23			
24	*Observation*	*Predicted Dreams*	*Residuals*
25	1	15.1	5.9
26	2	15.1	2.9
27	3	15.1	-1.1
28	4	15.1	4.9
29	5	15.1	-4.1
30	6	15.1	3.9
31	7	15.1	-7.1
32	8	15.1	-3.1
33	9	15.1	-2.1
34	10	15.1	-0.1
35	11	9.777777778	2.222222222
36	12	9.777777778	4.222222222
37	13	9.777777778	0.222222222
38	14	9.777777778	-1.777777778
39	15	9.777777778	6.222222222
40	16	9.777777778	-4.777777778
41	17	9.777777778	-6.777777778
42	18	9.777777778	-0.777777778
43	19	9.777777778	1.222222222

Interpreting the Output

Regression Statistics

- **Multiple R**. The square root of the ratio of SS Regression to SS Total, interpreted as the strength of the relationship between treatment and number of dreams.

$$R = \sqrt{\frac{SS_{Regression}}{SS_{Total}}} = \sqrt{\frac{134.1760}{434.6316}} = .5556$$

- **R Square**. The ratio of SS Regression to SS Total, interpreted as the proportion of variability in dreams that is explained by treatment.

$$R^2 = \frac{SS_{Regression}}{SS_{Total}} = \frac{134.1760}{434.6316} = .3087$$

- **Adjusted R Square**. R square adjusted to provide a more accurate estimate of the population R^2 value.

- **Standard Error**. Standard error of estimate, interpreted as the average error in predicting number of dreams by means of the regression equation.

$$SE_{\hat{Y}} = \sqrt{S_Y^2 (1 - R_{Adj}^2)} = \sqrt{\frac{SS_{Total}}{df_{Total}}(1 - R_{Adj}^2)} = \sqrt{\frac{434.6316}{18}(1 - .2680)} = 4.2041$$

- **Observations**. The number of observations included in the analysis.

ANOVA

- **Regression**. Variability in number of dreams that is explained by treatment condition.

- **Residual**. Variability in number of dreams that is not explained by treatment condition.

- **Total**. Total variability in number of dreams.

- **F**. The F value is obtained by dividing MS Regression by MS Residual. Note that the square root of F is equal to the t-test value obtained for these same data on page 172.

$$\sqrt{F} = \sqrt{7.5918} = 2.7553$$

- **Significance F**. The probability associated with the obtained value of F. Note that this probability (.0135) is identical to the two-tailed probability of obtained t displayed in the t-test output on page 172.

Coefficients

- **Intercept**. The intercept of the regression equation is equal to the mean of treatment group K, in this case, treatment group 2, the Placebo condition.

$$\overline{Y}_2 = 9.7778$$

- **X1**. The X1 coefficient is equal to the difference between the treatment group means.

$$\overline{Y}_1 - \overline{Y}_2 = 15.1 - 9.7778 = 5.3222$$

The regression equation for predicting number of dreams from treatment is

$$\hat{Y} = 9.7778 + 5.3222 X1$$

Residual Output

• **Predicted Dreams**. The predicted number of dreams for each subject is equal to the mean of the subject's treatment condition.

• **Residuals**. The residual for each subject is equal to that subject's observed number of dreams minus that subject's treatment condition mean.

Sample Research Problem with Three Groups

For an example of a three-group dummy coding problem, I will use the same data that were used for the one-way ANOVA on page 186. The regression analysis outcome will be identical to that of the one-way ANOVA.

Steps to Follow to Analyze the Sample Research Problem

1. Enter the data in an Excel worksheet as shown below.

	A	B
1	Group	Rating
2	Clean	8
3	Clean	5
4	Clean	7
5	Clean	6
6	Clean	3
7	Clean	7
8	Criminal	1
9	Criminal	2
10	Criminal	1
11	Criminal	3
12	Criminal	2
13	Control	4
14	Control	3
15	Control	5
16	Control	6
17	Control	4

2. Because treatment has three levels, two dummy variables can be defined. Let's call those variables X1 and X2 and place them in columns C and D, respectively.

The X1 and X2 values can be keyed in directly or you can generate them by using IF statements. For the X1 values, activate cell C2 and enter **=IF(A2="Clean",1,0)**. *Press* **[Enter]**. *Then copy the contents of cell C2 to cells C3 through C17. For the X2 values, activate cell D2 and enter* **=IF(A2="Criminal",1,0)**. *Press* **[Enter]**. *Then copy the contents of cell D2 to cells D3 through D17.*

	A	B	C	D
1	Group	Rating	X1	X2
2	Clean	6	1	0
3	Clean	5	1	0
4	Clean	7	1	0
5	Clean	6	1	0
6	Clean	3	1	0
7	Clean	7	1	0
8	Criminal	1	0	1
9	Criminal	2	0	1
10	Criminal	1	0	1
11	Criminal	3	0	1
12	Criminal	2	0	1
13	Control	4	0	0
14	Control	3	0	0
15	Control	5	0	0
16	Control	6	0	0
17	Control	4	0	0

3. In the Menu Bar, click on **Tools** and select **Data Analysis**.

If Data Analysis does not appear as a choice in the Tools menu, you will need to load the Microsoft Excel ToolPak add-in. Follow the procedure on page 7.

---

4. In the Analysis Tools window, select **Regression** and click **OK**.

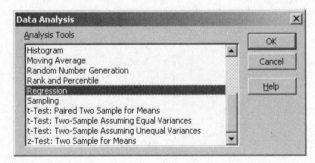

5. Complete the Regression dialog box as shown below. Detailed instructions are given immediately following the dialog box.

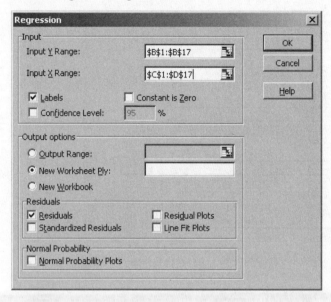

- **Input Y Range**. Enter the range of the Rating variable in the **Input Y Range** window. First click in the window. Then click in cell **B1** of the worksheet and drag to cell **B17**. If you prefer, you can manually enter **B1:B17**.
- **Input X Range**. Enter the range of the X1 and X2 variables in the **Input X Range** window. First click in the window. Then click in cell **C1** in the worksheet and drag to cell **D17**. Or, if you prefer, you can manually enter **C1:D17**.
- **Labels**. Click in the box next to **Labels** to place a check mark there. Cells B1, C1, and D1 contain the labels "Rating," "X1," and "X2," respectively, and should not be included in the regression analysis. They will, however, be used to label the output.
- **New Worksheet Ply**. Click in the circle next to **New Worksheet Ply** to place the output in a sheet different from the sheet containing the data.
- **Residuals**. Click in the box next to **Residuals**.
6. Click **OK**. The output that you will obtain is shown on the next page.

	A	B	C	D	E	F	G
1	SUMMARY OUTPUT						
2							
3	*Regression Statistics*						
4	Multiple R	0.826277824					
5	R Square	0.682735043					
6	Adjusted R Square	0.633925049					
7	Standard Error	1.219499687					
8	Observations	16					
9							
10	ANOVA						
11		*df*	*SS*	*MS*	*F*	*Significance F*	
12	Regression	2	41.60416667	20.80208	13.98761	0.000574439	
13	Residual	13	19.33333333	1.487179			
14	Total	15	60.9375				
15							
16		*Coefficients*	*Standard Error*	*t Stat*	*P-value*	*Lower 95%*	*Upper 95%*
17	Intercept	4.4	0.54537684	8.067816	2.04E-06	3.221784972	5.578215
18	X1	1.266666667	0.738443732	1.715319	0.110008	-0.32864402	2.8619774
19	X2	-2.6	0.771279324	-3.37102	0.005014	-4.26624767	-0.933752
20							

	A	B	C	D	E	F	G
22							
23	RESIDUAL OUTPUT						
24							
25	*Observation*	*Predicted Rating*	*Residuals*				
26	1	5.666666667	0.333333333				
27	2	5.666666667	-0.66666667				
28	3	5.666666667	1.333333333				
29	4	5.666666667	0.333333333				
30	5	5.666666667	-2.66666667				
31	6	5.666666667	1.333333333				
32	7	1.8	-0.8				
33	8	1.8	0.2				
34	9	1.8	-0.8				
35	10	1.8	1.2				
36	11	1.8	0.2				
37	12	4.4	-0.4				
38	13	4.4	-1.4				
39	14	4.4	0.6				
40	15	4.4	1.6				
41	16	4.4	-0.4				

Interpreting the Output

Regression Statistics

- **Multiple R**. The square root of the ratio of SS Regression to SS Total, interpreted as the strength of the relationship between treatment and rating.

$$R = \sqrt{\frac{SS_{Regression}}{SS_{Total}}} = \sqrt{\frac{41.6042}{60.9375}} = .8263$$

- **R Square**. The ratio of SS Regression to SS Total, interpreted as the proportion of variability in ratings that is explained by treatment.

$$R^2 = \frac{SS_{Regression}}{SS_{Total}} = \frac{41.6042}{60.9375} = .6827$$

- **Adjusted R Square**. R square adjusted to provide a more accurate estimate of the population R^2 value.

- **Standard Error**. Standard error of estimate, interpreted as the average error in predicting the ratings by means of the regression equation.

$$SE_{\hat{Y}} = \sqrt{S_Y^2(1 - R_{Adj}^2)} = \sqrt{\frac{SS_{Total}}{df_{Total}}(1 - R_{Adj}^2)} = \sqrt{\frac{60.9375}{15}(1 - .6339)} = 1.2195$$

- **Observations**. The number of observations included in the analysis.

ANOVA

- **Regression**. Variability in the ratings that is explained by treatment condition.

- **Residual**. Variability in the ratings that is not explained by treatment condition.

- **Total**. Total variability in the ratings.

- **F**. The F value is obtained by dividing MS Regression by MS Residual. Note that this F of 13.9876 is equal to the one-way ANOVA F-test value obtained for these same data on page 187.

- **Significance F**. The probability associated with the obtained value of F. Note that this probability (.0005) is identical to the probability of obtained F displayed in the ANOVA output on page 187.

Coefficients

- **Intercept**. The intercept of the regression equation is equal to the mean of treatment group K, in this case, group 3, the Control treatment group.

$$\overline{Y}_3 = 4.4$$

- **X1**. The X1 coefficient is equal to the difference between the means of group 1 (Clean) and group 3 (Control).

$$\overline{Y}_1 - \overline{Y}_3 = 5.6667 - 4.4 = 1.2667$$

- **X2**. The X2 coefficient is equal to the difference between the means of group 2 (Criminal) and group 3 (Control).

$$\overline{Y}_2 - \overline{Y}_3 = 1.8 - 4.4 = -2.6$$

The regression equation for predicting rating from treatment is given by

$$\hat{Y} = 4.4 + 1.2667X1 + (-2.6)X2$$

Residual Output

- **Predicted Rating**. The predicted rating for each subject is equal to the mean of the subject's treatment condition.

- **Residuals**. The residual for each subject is equal to that subject's rating minus that subject's treatment condition mean.

Curvilinear Regression

When the relationship between a predictor variable (X) and the outcome variable (Y) is nonlinear, you may want to carry out a regression analysis to investigate a polynomial relationship between X and Y. In this section, I explain how to use Excel's procedures to apply the model shown below.

$$\hat{Y} = B_0 + B_1 X + B_2 X^2$$

Sample Research Problem

A statistics instructor was interested in finding out if there was a curvilinear relationship between statistics anxiety and performance on the statistics final exam. The instructor recorded the final exam score and an anxiety measure for each student enrolled the class.

Steps to Follow to Analyze the Sample Research Problem

1. Enter the data in an Excel worksheet.

	A	B
1	Final Exam	Anxiety
2	19	5
3	18	7
4	21	10
5	23	12
6	28	15
7	38	16
8	37	17
9	36	16
10	20	18
11	16	19
12	14	20
13	8	21

2. For this example, Final Exam is the Y variable and Anxiety is the X variable. We need to square the X variable. Type the label **Anxiety Squared** in cell **C1**.

	A	B	C	D
1	Final Exam	Anxiety	Anxiety Squared	
2	19	5		
3	18	7		

3. Click in cell **C2** and enter the formula **=B2^2**. Press [**Enter**].

	A	B	C	D
1	Final Exam	Anxiety	Anxiety Squared	
2	19	5	=B2^2	

4. Copy the formula in cell C2 to cells C3 through C13.

	A	B	C	D
1	Final Exam	Anxiety	Anxiety Squared	
2	19	5	25	
3	18	7	49	
4	21	10	100	
5	23	12	144	
6	28	15	225	
7	38	16	256	
8	37	17	289	
9	36	16	256	
10	20	18	324	
11	16	19	361	
12	14	20	400	
13	8	21	441	

5. Click **Tools** in the Menu Bar and select **Data Analysis**.

If Data Analysis does not appear as a choice in the Tools menu, you will need to load the Microsoft Excel ToolPak add-in. Follow the procedure on page 7.

6. In the Data Analysis dialog box, select **Regression** and click **OK**.

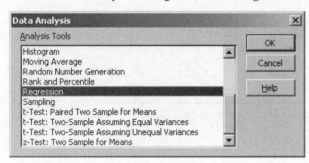

7. Complete the Regression dialog box as shown below.

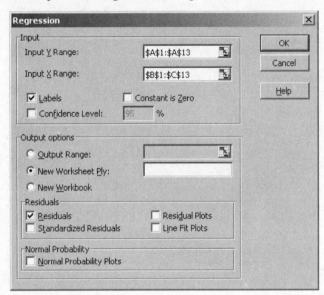

8. Click **OK**. The summary output provided by Excel's regression procedure is shown below and at the top of the next page.

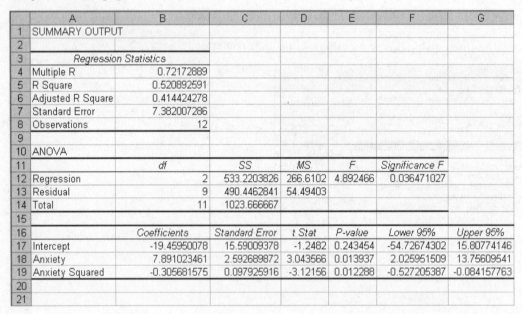

	A	B	C	D	E	F	G
1	SUMMARY OUTPUT						
2							
3	*Regression Statistics*						
4	Multiple R	0.72172889					
5	R Square	0.520892591					
6	Adjusted R Square	0.414424278					
7	Standard Error	7.382007286					
8	Observations	12					
9							
10	ANOVA						
11		*df*	*SS*	*MS*	*F*	*Significance F*	
12	Regression	2	533.2203826	266.6102	4.892466	0.036471027	
13	Residual	9	490.4462841	54.49403			
14	Total	11	1023.666667				
15							
16		*Coefficients*	*Standard Error*	*t Stat*	*P-value*	*Lower 95%*	*Upper 95%*
17	Intercept	-19.45950078	15.59009378	-1.2482	0.243454	-54.72674302	15.80774146
18	Anxiety	7.891023461	2.592689872	3.043566	0.013937	2.025951509	13.75609541
19	Anxiety Squared	-0.305681575	0.097925916	-3.12156	0.012288	-0.527205387	-0.084157763
20							
21							

		Predicted Final Exam	Residuals				
22							
23	RESIDUAL OUTPUT						
24							
25	Observation	Predicted Final Exam	Residuals				
26	1	12.35357715	6.646422848				
27	2	20.79926628	-2.799266277				
28	3	28.88257634	-7.882576344				
29	4	31.21463397	-8.214633974				
30	5	30.1274968	-2.127496797				
31	6	28.54239144	9.457608562				
32	7	26.34592293	10.65407707				
33	8	28.54239144	7.457608562				
34	9	23.53809127	-3.538091273				
35	10	20.11889647	-4.118896466				
36	11	16.08833851	-2.088338509				
37	12	11.4464174	-3.446417403				

Interpreting the Output

Regression Statistics.

- **Multiple R**. The multiple correlation, R, between the two predictor variables (Anxiety and Anxiety-squared) and Y (Final Exam) is .7217.

- **R Square**. R^2 is equal to .5209. Approximately 52 percent of the variation in Final Exam is explainable by the polynomial regression model.

- **Adjusted R Square**. This is an adjusted value that more closely reflects the degree of fit in the population. For this sample problem, adjusted R^2 is equal to .4144.

- **Standard Error**. The standard error of estimate is equal to 7.3820. This is the average amount by which we err if we use our polynomial regression equation to predict Final Exam.

- **Observations**. The data set includes data for 12 statistics students.

ANOVA

- **Regression**. The variation in Final Exam scores that is explained by its relationship with the predictor variables is referred to as *regression*.

- **Residual**. The variation in Final Exam scores that is not explained by the predictor variables is referred to as *residual*.

- **Total**. The total variation in Final Exam scores is partitioned into regression variation and residual variation.

The F value is obtained by dividing MS regression by MS residual. This F test value would be considered statistically significant with alpha equal to .05, because the P-value (.0365) is less than .05.

Coefficient Summary Information

- **Intercept**. The intercept of the regression equation is –19.4595.

- **Anxiety beta weight**. The beta weight for the Anxiety score is 7.8910.

- **Anxiety-squared beta weight**. The beta weight for the curvilinear component, the squared Anxiety score, is –0.3057.

When these values are placed in the regression equation, we see that the polynomial equation for predicting Final Exam from Anxiety is given by

$$\hat{Y} = -19.4595 + 7.8910X - 0.3057X^2$$

If we use the data for the first person in the data set, we obtain

Predicted Final Exam = $-19.4595 + 7.8910(5) - 0.3057(5^2) = 12.35$

- The **Standard Error** values, **t Stat**, **P-value**, and **Lower** and **Upper 95%** confidence interval boundaries are also provided for each of these coefficients. The t Stat for the intercept provides a statistical test of the hypothesis that the intercept is equal to zero. The t tests for the predictor variable beta weights both test the hypothesis that the beta weight is equal to zero. Of special interest here, is the t test for the beta weight of the squared Anxiety measure, because it is this beta weight that reflects the curvilinear component. For this example, the curvilinear component is statistically significant with a P-value equal to .0123.

Residuals

- **Observation**.   Refers to the order in which the observations appear in the data set. For example, observation 1 is the first observation in the data set, observation 2 is the second observation in the data set, and so on.

- **Predicted Final Exam**. The value we obtain when we use the polynomial regression equation to predict a student's Final Exam score. The predicted Final Exam score for observation 1 is 12.35.

- **Residuals**. Residuals found by subtracting the predicted value from the actual value. For observation 1, $19 - 12.3536 = 6.6464$.

Scatterplot of a Curvilinear Relation

Excel's Chart Wizard enables us to produce a scatterplot with a trendline and also to show the regression equation and R^2 on the graph.

1. Return to the worksheet with the data for the curvilinear regression example problem. Arrange the columns of data so that they appear exactly as shown below. Then click in any cell containing data.

	A	B	C
1	Anxiety	Final Exam	
2	5	19	
3	7	18	
4	10	21	
5	12	23	
6	15	28	
7	16	38	
8	17	37	
9	16	36	
10	18	20	
11	19	16	
12	20	14	
13	21	8	

2. Click **Insert** in the Menu Bar and select **Chart**.

3. **Chart Wizard-Step 1 of 4-Chart Type**. Select **XY(Scatter)** and click **Next>**.

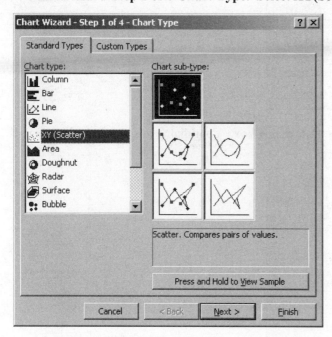

4. **Source Data**. If the correct data range does not show in the Data range window, click in the window and enter the information exactly as it appears below (=Sheet1!$A$1:$B$13).

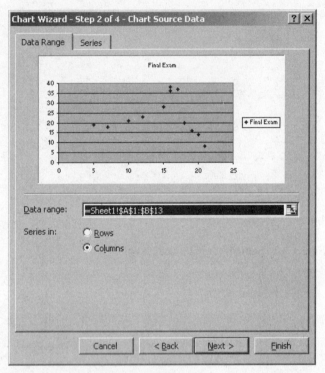

5. Click the **Series** tab at the top of the Source Data dialog box. Your dialog box should have the same entries as the one shown below. Click **Next>**.

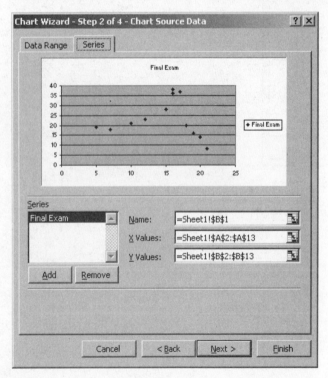

6. Click the **Titles** tab at the top of the Chart Options dialog box. Key in the tiles as shown below.

Chart title: **Prediction of Exam Performance from Anxiety**

Value (X) axis: **Anxiety**

Value (Y) axis: **Final Exam Score**

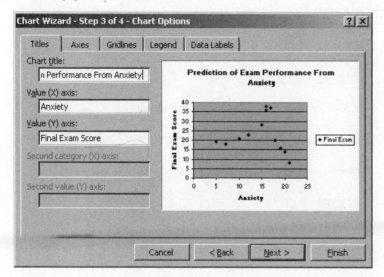

7. Click the **Legend** tab at the top of the dialog box. Click in the box to the left of **Show legend** to remove the checkmark that appears there. Click **Next>**.

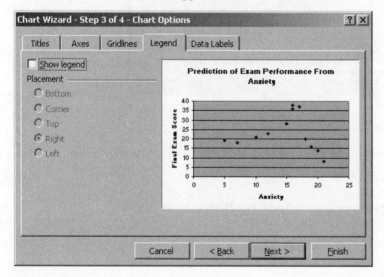

8. Let's place the scatterplot in a new worksheet. Click the button to the left of **As new sheet**. Click **Finish**.

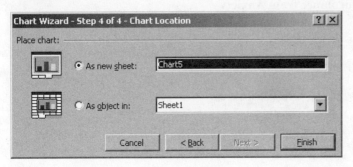

9. Right-click directly on one of the dots. Select **Add Trendline** from the menu.

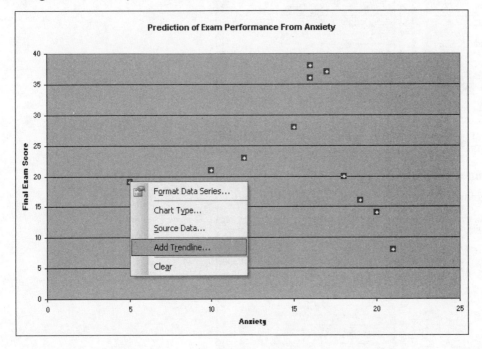

10. We will place two trendlines on the graph: One for the linear relationship, and one for the curvilinear relationship. Select **Linear** and click **OK**.

11. Right-click directly on one of the dots and select **Add Trendline** from the menu.

12.   Select the **Polynomial** type of Order **2**. Click the **Options** tab at the top of the dialog box.

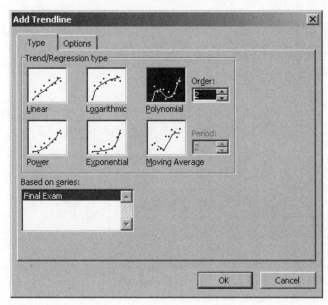

13.   Click in the box to the left of **Display equation on chart**. Click in the box to the left of **Display R-squared value on chart**. Click **OK**.

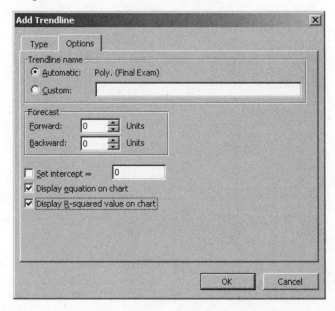

14.   Click and drag the output so that it is placed where it can be easily read.

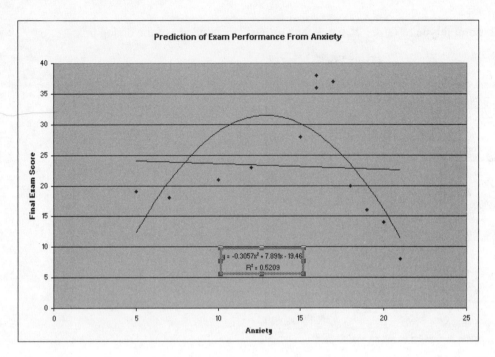

15. Let's increase the font size of the regression equation so that it is easier to read. Click on the equation so that it has a grey border.

16. Click **Format** in the Menu Bar and select **Selected Data Labels**.

17. Click the **Font** tab at the top of the Format Data Labels dialog box. Select font size **14**. Click **OK**. The completed graph is shown below.

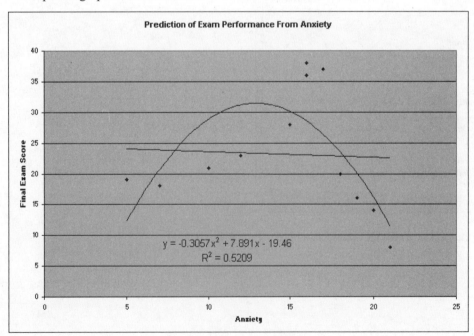

Cross Tabulations Using the Pivot Table

Excel's Pivot Table provides an easy way to cross tabulate both qualitative and quantitative variables. In this chapter, I present three different types of cross tabulation examples: 1) two qualitative variables, 2) one qualitative and one quantitative variable, and 3) three variables (two qualitative and one quantitative). In addition, I show how missing values are displayed.

Sample Research Problem

A survey research project was carried out to gather information about people's attitudes regarding capital punishment. "For" was recorded if respondents indicated they were in favor of capital punishment, and "Against" was recorded if respondents indicated they were against capital punishment. The survey instrument also included some items regarding the participants' demographic characteristics: sex, socioeconomic status (SES), and age. One of the participants did not provide a response for SES, so that cell was left blank. The worksheet containing the data is shown below.

	A	B	C	D
1	OPINION	SEX	SES	AGE
2	Against	Male	Hi	31
3	For	Male	Med	39
4	Against	Female	Med	40
5	Against	Male	Med	51
6	For	Female	Med	40
7	For	Female	Med	42
8	For	Female		40
9	Against	Male	Med	41
10	For	Female	Lo	31
11	Against	Male	Lo	39
12	Against	Female	Lo	54
13	For	Female	Hi	51
14	Against	Male	Med	39
15	Against	Female	Hi	40
16	For	Male	Hi	55
17	For	Male	Lo	32
18	Against	Female	Med	39

Cross Tabulation of Two Qualitative Variables

To illustrate how to cross tabulate two qualitative variables, we will work with OPINION regarding capital punishment and SEX of the respondent.

1. First enter the data in an Excel worksheet as shown above. Then click **Data** in the Menu Bar and select **Pivot Table and Pivot Chart Report**.

2. **Pivot Table and Pivot Chart Wizard-Step 1 of 3**. The data are located in an Excel worksheet, so select **Microsoft Excel List or Database**. You want to create a Pivot Table report, so select **Pivot Table**. Click **Next>**.

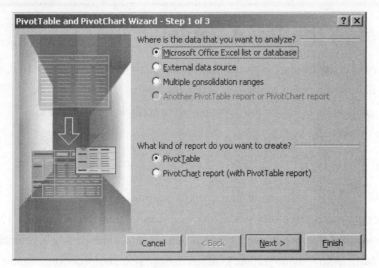

3. **Pivot Table and Pivot Chart Wizard-Step 2 of 3**. Enter the range of the data (A1 to D18) in the space provided. To do this, first click in the **Range** window. Then click on cell **A1** in the worksheet and drag to cell **D18**. If you prefer, you can manually enter **A1:D18**. Click **Next>**.

4. **Pivot Table and Pivot Chart Wizard-Step 3 of 3**. Let's put the report in the same worksheet as the data. Select **Existing worksheet**. Click in the **Existing worksheet** window and then click on cell **A20** in the worksheet to place the report a couple lines below the data, with cell A20 as the uppermost left cell of the output. If you prefer, you can manually enter **A20** in the window. Click **Finish**.

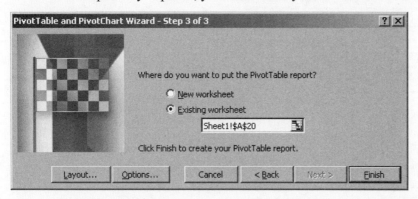

5. Depending on your preferences, you can place OPINION on either the rows or the columns of the cross tabulation. For this example, we will place OPINION on the columns and SEX on the rows. Then, to obtain a count of OPINION responses cross tabulated by SEX, we will place OPINION on the data area of the table. In the Pivot Table Field List, click on **OPINION** to select it. Next, at the bottom of the Pivot Table Field List, select **Column Area**. Then, click the **Add To** button.

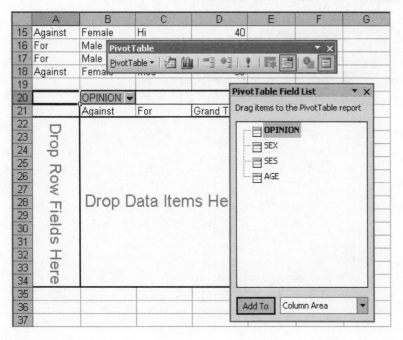

6.   Click on **SEX** to select it. Select **Row Area**. Click the **Add To** button.

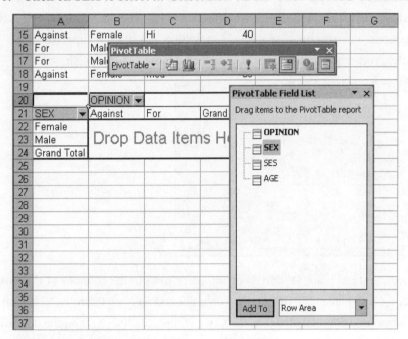

7.   Click on **OPINION** to select it. Select **Data Area**. Click the **Add To** button.

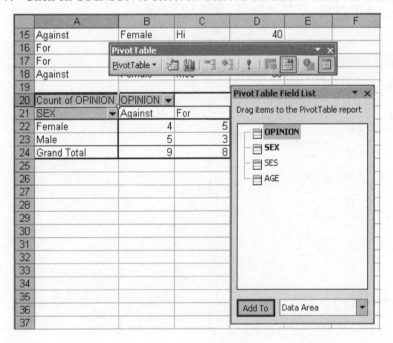

8. Close the Pivot Table Field List dialog box. Excel automatically displays labels in alphabetical order.
    Let's change the order of the row labels and put Male first and Female second. Click in cell **A23** where
    Male appears. Click **Pivot Table** and select **Order→Move Up** from the drop down menu.

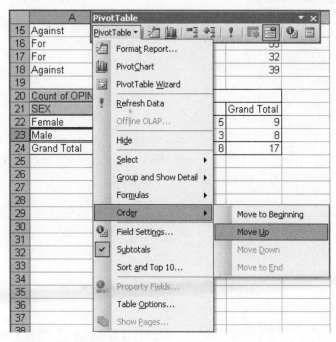

9. Let's also change the order of the column labels. Let's put For first and Against second. Click in cell
    B21 where Against appears. Click **Pivot Table** and select **Order→Move Right** from the drop down
    menu.

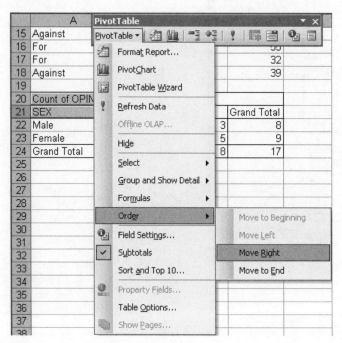

10.   Let's use a more professional table format. Click any cell within the table. Then click **Pivot Table** and select **Format Report**.

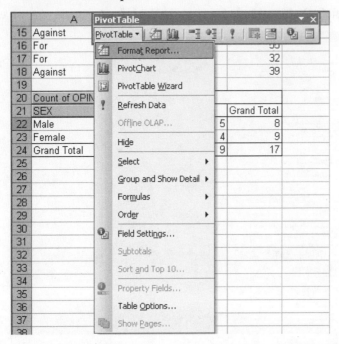

11.   Ten different format options are available. For this example, click on **Report 4** to select it. Click **OK**.

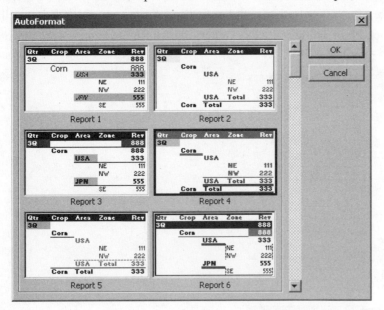

The completed table is shown below.

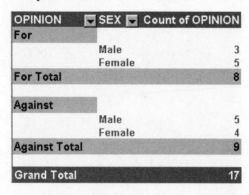

OPINION	SEX	Count of OPINION
For		
	Male	3
	Female	5
For Total		8
Against		
	Male	5
	Female	4
Against Total		9
Grand Total		17

Cross Tabulation of a Qualitative and a Quantitative Variable

For my illustration of cross tabulating a qualitative variable with a quantitative variable, I will utilize the responses given to **OPINION** and **AGE**.

Before beginning this example, the data must be entered in an Excel worksheet. Enter the data on page 259 if you haven't already done so.

1. The worksheet containing the data (most likely Sheet 1) should be displayed on your screen. If it is not, return to that worksheet now. To do this, click the Sheet 1 tab near the bottom of the screen.

	A	B	C	D
1	OPINION	SEX	SES	AGE
2	Against	Male	Hi	31
3	For	Male	Med	39
4	Against	Female	Med	40
5	Against	Male	Med	51
6	For	Female	Med	40
7	For	Female	Med	42
8	For	Female		40
9	Against	Male	Med	41
10	For	Female	Lo	31
11	Against	Male	Lo	39
12	Against	Female	Lo	54
13	For	Female	Hi	51
14	Against	Male	Med	39
15	Against	Female	Hi	40
16	For	Male	Hi	55
17	For	Male	Lo	32
18	Against	Female	Med	39

2. Click **Data** in the Menu Bar and select **Pivot Table and Pivot Chart Report**. Or, if the Pivot Table toolbar is displayed, click **Pivot Table** and select **Pivot Table Wizard**.

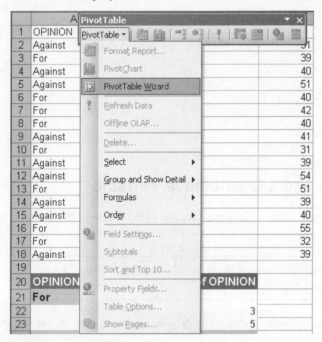

3. **Pivot Table and Pivot Chart Wizard-Step 1 of 3**. The data are located in an Excel worksheet, so select **Microsoft Excel List or Database**. You want to generate a Pivot Table report, so select **Pivot Table**. Click **Next>**.

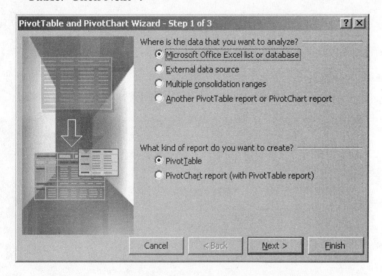

4. **Pivot Table and Pivot Chart Wizard-Step 2 of 3**. Enter the range of the data (A1 to D18). To do this, first click in the **Range** window. Then either click in cell **A1** in the worksheet and drag to cell **D18**, or manually enter **A1:D18**. Click **Next>**.

---

*You might see a message that starts out "Your new report will use less memory if you base it on your existing report . . ." Unless memory is a problem, I usually click **No**.*

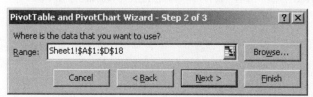

5. **Pivot Table and Pivot Chart Wizard-Step 3 of 3**. Let's put the report in a new worksheet. Select **New worksheet** and, at the bottom of the dialog box, click **Layout**.

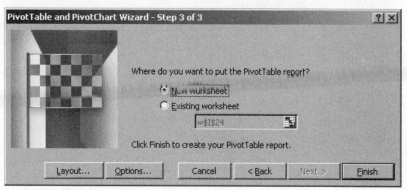

6.   OPINION is recorded as either "For" or "Against." AGE, however, takes on several values from 31 to 55. Therefore, the table will be more attractive and will more easily fit on one page if you place AGE on the rows and OPINION on the columns. Drag the **OPINION** field button to COLUMN. Drag the **AGE** field button to ROW. When you are working with two qualitative variables, it does not matter a great deal which one you drag to the data area of the Pivot Table. Quantitative variables, however, may require an extra step that you should be aware of. So, drag **AGE** (a quantitative variable in this example) to the space labeled **DATA**.

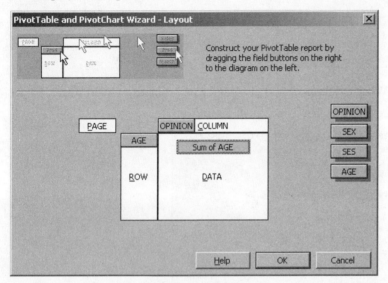

7.   You see **Sum of AGE** in the data area because the default summary measure for a quantitative variable is sum. You want count of age, not sum of age. To change this, double click on **Sum of Age**.

8.   In the **Summarize by** window, select **Count** and click **OK**. .

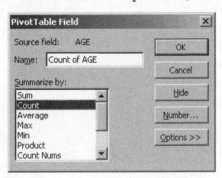

9. Click **OK** in the Layout dialog box. Click **Finish** in the Step 3 of 3 dialog box. The table shown below now displays counts rather than sums. The table is not quite finished, however, because there are a number of empty cells. You would like to see zeroes in the empty cells.

	A	B	C	D
1	Drop Page Fields Here			
2				
3	Count of AGE	OPINION ▼		
4	AGE ▼	Against	For	Grand Total
5	31	1	1	2
6	32		1	1
7	39	3	1	4
8	40	2	2	4
9	41	1		1
10	42		1	1
11	51	1	1	2
12	54	1		1
13	55		1	1
14	Grand Total	9	8	17

10. Right-click on a cell within the table. Select **Table Options** from the drop down menu that appears.

11. Click in the box next to **For empty cells, show:** to remove the check mark. Click **OK**.

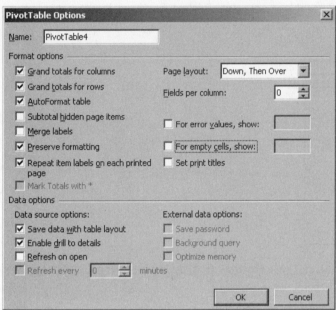

The final version of the table is shown below.

	A	B	C	D
1				
2				
3	Count of AGE	OPINION ▾		
4	AGE ▾	Against	For	Grand Total
5	31	1	1	2
6	32	0	1	1
7	39	3	1	4
8	40	2	2	4
9	41	1	0	1
10	42	0	1	1
11	51	1	1	2
12	54	1	0	1
13	55	0	1	1
14	Grand Total	9	8	17

Cross Tabulation of Three Variables

For a cross tabulation example with three variables, we will use OPINION, SEX, and AGE. Because OPINION and SEX both take on only two different values and AGE takes on several values, we will place AGE on the rows of the table and both OPINION and SEX on the columns. This layout generates a table that is easier to read and use than if we request the opposite placement of the variables on the rows and columns of the table. I suggest that you experiment with different layouts to find the presentation that best fits your needs.

It really does not matter if you place OPINION or SEX in the data area of the table. What does matter is the order in which OPINION and SEX are placed in the column dimension of the table. I will show the output that is obtained with the two different orders.

Before beginning this example, the data must be entered in an Excel worksheet. Enter the
data on page 259 if you haven't already done so.

1. First return to the worksheet that contains the data (most likely Sheet 1). To do this, click the Sheet 1
    tab near the bottom of the worksheet.

	A	B	C	D
1	OPINION	SEX	SES	AGE
2	Against	Male	Hi	31
3	For	Male	Med	39
4	Against	Female	Med	40
5	Against	Male	Med	51
6	For	Female	Med	40
7	For	Female	Med	42
8	For	Female		40
9	Against	Male	Med	41
10	For	Female	Lo	31
11	Against	Male	Lo	39
12	Against	Female	Lo	54
13	For	Female	Hi	51
14	Against	Male	Med	39
15	Against	Female	Hi	40
16	For	Male	Hi	55
17	For	Male	Lo	32
18	Against	Female	Med	39

2. Click the **Data** button in the Menu Bar, and select **Pivot Table and Pivot Chart Report**. Or, if the
    Pivot Table toolbar is displayed, click **Pivot Table** and select **Pivot Table Wizard**.

3. **Pivot Table and Pivot Chart Wizard-Step 1 of 3**. The data are located in an Excel worksheet, so
    select **Microsoft Excel List or Database**. You want to create a Pivot Table report, so select **Pivot
    Table**. Click **Next>**.

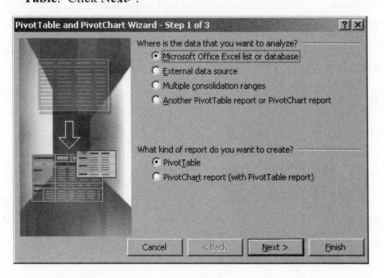

4. **Pivot Table and Pivot Chart Wizard-Step 2 of 3**. Enter the data range (A1 to D18) in the space provided. To do this, first click in the **Range** window. Then either click in cell **A1** in the worksheet and drag to cell **D18** or manually enter **A1:D18**. Click **Next>**.

    *You might see a message that starts out "Your new report will use less memory if you base it on your existing report . . . " Unless memory is a problem, I usually click **No**.*

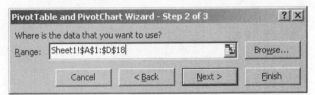

5. **Pivot Table Wizard-Step 3 of 3**. Let's place the table in a new worksheet. Select **New worksheet** and click **Finish**.

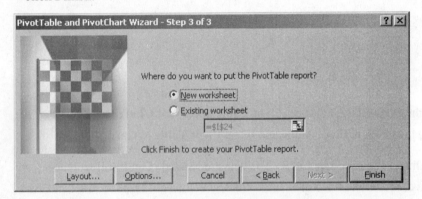

6. For this version of the table, we will place OPINION on the columns first, followed by SEX. In both versions, AGE will be placed on the rows, and OPINION will be placed in the data area. In the Pivot Table Field List, click on **OPINION** to select it. Next, at the bottom of the Pivot Table Field List, select **Column Area**. Then, click the **Add To** button.

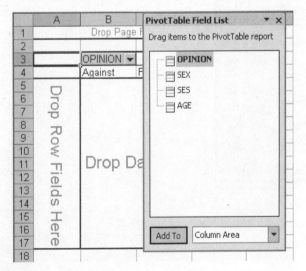

7. Click on **SEX** to select it. Select **Column Area**. Click the **Add To** button.

8. Click on **AGE** to select it. Select **Row Area**. Click the **Add To** button.

9. Click on **OPINION** to select it. Select **Data Area**. Click the **Add To** button.

10. If you would like to see zeroes in the blank cells, right-click in any cell of the table and select **Table Options** from the drop down menu that appears.

	A	B	C	D	E	F	G	H
1								
2								
3	Count of OPINION	OPINION ▼	SEX ▼					
4		Against		Against Total	For		For Total	Grand Total
5	AGE ▼	Female	Male		Female	Male		
6	31		1	1	1		1	2
7	32					1	1	1
8	39	1	2	3		1	1	4
9	40	2		2	2		2	4
10	41		1	1				1
11	42				1		1	1
12	51		1	1	1		1	2
13	54	1		1				1
14	55					1	1	1
15	Grand Total	4	5	9	5	3	8	17

11. Click in the box next to **For empty cells, show:** to remove the check mark. Click **OK**.

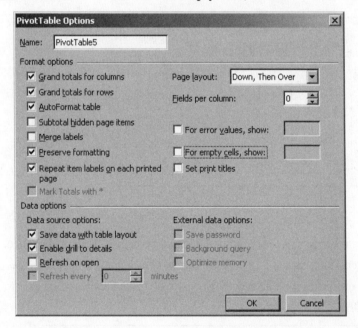

The completed table with zeroes displayed in empty cells is shown below.

	A	B	C	D	E	F	G	H
1								
2								
3	Count of OPINION	OPINION ▼	SEX ▼					
4		Against		Against Total	For		For Total	Grand Total
5	AGE ▼	Female	Male		Female	Male		
6	31	0	1	1	1	0	1	2
7	32	0	0	0	0	1	1	1
8	39	1	2	3	0	1	1	4
9	40	2	0	2	2	0	2	4
10	41	0	1	1	0	0	0	1
11	42	0	0	0	1	0	1	1
12	51	0	1	1	1	0	1	2
13	54	1	0	1	0	0	0	1
14	55	0	0	0	0	1	1	1
15	Grand Total	4	5	9	5	3	8	17

12. **Modifying the table**. Let's see what the table looks like when we change the order in which SEX and OPINION are placed in the layout. To make this modification, click on the **SEX** field button in cell C3 and drag it over to cell B3, on top of **OPINION**. Release the mouse key. The contents of the table will be re-tabulated with OPINION placed before SEX on the columns. Note that, from left to right in the column area of the table, the OPINION field button comes first and then SEX.

	A	B	C	D	E	F	G	H
1								
2								
3	Count of OPINION	SEX ▼	OPINION ▼					
4		Female		Female Total	Male		Male Total	Grand Total
5	AGE ▼	Against	For		Against	For		
6	31	0	1	1	1	0	1	2
7	32	0	0	0	0	1	1	1
8	39	1	0	1	2	1	3	4
9	40	2	2	4	0	0	0	4
10	41	0	0	0	1	0	1	1
11	42	0	1	1	0	0	0	1
12	51	0	1	1	1	0	1	2
13	54	1	0	1	0	0	0	1
14	55	0	0	0	0	1	1	1
15	Grand Total	4	5	9	5	3	8	17

Missing Data

When you use Excel for statistical analysis, missing values should be indicated by blanks in your worksheet. I have only one missing value in my data set which occurs in the SES column. The subject whose data are displayed in row 8 did not provide an SES response. To illustrate how the Pivot Table handles missing values, let's cross tabulate OPINION and SES.

Before beginning this example, the data must be entered in an Excel worksheet. Enter the data on page 259 if you haven't already done so.

1. Return to the sheet that contains the data set (most likely Sheet 1). To do this, click the Sheet 1 tab near the bottom of the screen.

	A	B	C	D
1	OPINION	SEX	SES	AGE
2	Against	Male	Hi	31
3	For	Male	Med	39
4	Against	Female	Med	40
5	Against	Male	Med	51
6	For	Female	Med	40
7	For	Female	Med	42
8	For	Female		40
9	Against	Male	Med	41
10	For	Female	Lo	31
11	Against	Male	Lo	39
12	Against	Female	Lo	54
13	For	Female	Hi	51
14	Against	Male	Med	39
15	Against	Female	Hi	40
16	For	Male	Hi	55
17	For	Male	Lo	32
18	Against	Female	Med	39

2. Click the **Data** button in the Menu Bar, and select **Pivot Table and Pivot Chart Report**.

3. **Pivot Table Wizard-Step 1 of 3**   Select **Microsoft Excel List or Database**. Select **Pivot Table**. Click **Next>**.

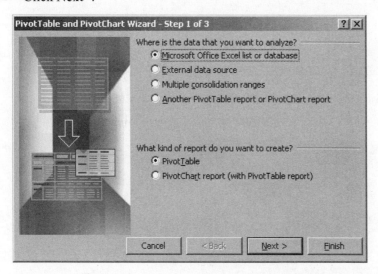

4. **Pivot Table and Pivot Chart Wizard-Step 2 of 3**. Enter the data range (A1:D18) in the space provided. To do this, first click in the **Range** window. Then click in cell **A1** of the worksheet and drag to cell **D18**. If you prefer, you can manually enter **A1:D18**. Click **Next>**.

---

*You might see a message that starts out "Your new report will use less memory if you base it on your existing report . . . " Unless memory is a problem, I usually click **No**.*

5. **Pivot Table and Pivot Chart Wizard-Step 3 of 3**. Let's put the report in a new worksheet. Select **New worksheet** and click **Layout** at the bottom of the dialog box.

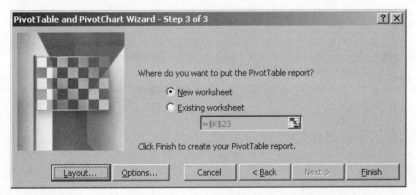

6. Drag the **SES** field button to COLUMN. Drag the **OPINION** field button to ROW. Drag the **OPINION** field button to DATA. Click **OK**. Click **Finish** in the Step 3 of 3 dialog box.

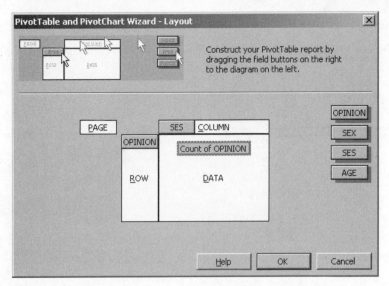

The completed cross tabulation is shown below. Notice the column headed "(blank)". This column contains a tally of the OPINION responses for the survey participants who did not answer the SES question.

	A	B	C	D	E	F
1						
2						
3	Count of OPINION	SES ▼				
4	OPINION ▼	Hi	Lo	Med	(blank)	Grand Total
5	Against	2	2	5		9
6	For	2	2	3	1	8
7	Grand Total	4	4	8	1	17

Chi-Square Test of Independence

When a researcher is interested in investigating the relationship between qualitative variables, the appropriate test is often the chi-square test of independence. To carry out this test, we first need to calculate the observed and expected frequencies for each cell of the cross tabulation. Although the Pivot Table can be used to provide observed frequencies, we will need to key in our own formulas to calculate the expected frequencies. Once we have the observed and expected frequencies, Excel's functions can be utilized to complete the computations for chi-square and to evaluate its statistical significance.

The formula for the chi-square test of independence is given by

$$X^2 = \sum \frac{(O_j - E_j)^2}{E_j}$$, where O_j is an observed cell frequency and E_j is an expected cell frequency.

Assumptions Underlying the Chi-Square Test of Independence

Four assumptions underlie the chi-square test of independence:

1. Subjects are randomly and independently sampled from the population of interest.
2. Measures are obtained from a single sample.
3. Variables included in the analysis are measured on a qualitative scale.
4. Expected cell frequencies are greater than or equal to five.

Sample Research Problem

A researcher had a hunch that people who did not smoke would be more likely to select a marital partner who also did not smoke rather than someone who did. The researcher administered a questionnaire to a sample of 22 married couples who indicated whether they or their spouses were smokers. Smoking behavior was categorized as either "smoker" or "nonsmoker."

Using the Pivot Table for Observed Cell Frequencies

1. First enter the data in an Excel worksheet as displayed at the top of the next page.

	A	B
1	Husband	Wife
2	Smoker	Smoker
3	Smoker	Nonsmoker
4	Nonsmoker	Nonsmoker
5	Smoker	Nonsmoker
6	Nonsmoker	Nonsmoker
7	Smoker	Smoker
8	Nonsmoker	Nonsmoker
9	Nonsmoker	Smoker
10	Smoker	Smoker
11	Nonsmoker	Nonsmoker
12	Smoker	Smoker
13	Smoker	Nonsmoker
14	Nonsmoker	Nonsmoker
15	Smoker	Smoker
16	Smoker	Smoker
17	Nonsmoker	Nonsmoker
18	Nonsmoker	Nonsmoker
19	Smoker	Smoker
20	Smoker	Nonsmoker
21	Nonsmoker	Nonsmoker
22	Nonsmoker	Nonsmoker
23	Nonsmoker	Smoker

2. Click on **Data** in the Menu Bar and select **Pivot Table and Pivot Chart Report**.

3. **Pivot Table and Pivot Chart Wizard-Step 1 of 3**. Because the data are in an Excel worksheet, select **Microsoft Excel List or Database**. You want to create a Pivot Table report, so select **Pivot Table**. Click **Next>**.

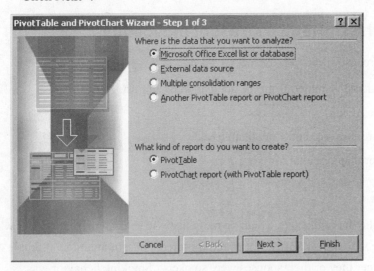

4. **Pivot Table and Pivot Chart Wizard-Step 2 of 3**. Fill in the range for the data set as shown below. The data are recorded in cells A1 through B23. To enter the data range, first click in the **Range** window. Then click in cell **A1** in the worksheet and drag to cell **B23**. Or, if you prefer, you can manually enter **A1:B23** in the range window. Click **Next>**.

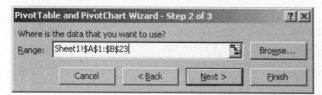

5. **Pivot Table and Pivot Chart Wizard-Step 3 of 3**. Let's put the report in the same worksheet as the data. Select **Existing worksheet**. Click in the **Existing worksheet** window and then click in cell **D1** of the worksheet to place the report to the right of the data with D1 as the uppermost left cell. If you prefer, you can manually enter **D1** in the Existing worksheet window. Click **Layout** at the bottom of the dialog box.

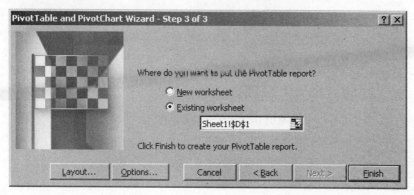

6. Drag the **Husband** field button to COLUMN. Drag the **Wife** field button to ROW. Drag the **Husband** field button to DATA.

When you are cross tabulating just two variables, it does not matter which one is placed in the data area of the table.

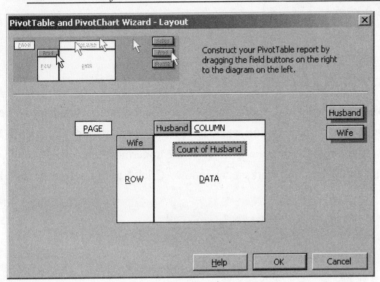

7. Click **OK**. Click **Finish** in the Step 3 of 3 dialog box.   The completed table is shown below in D1:G5. This table contains the observed frequencies that will be used to calculate chi-square.

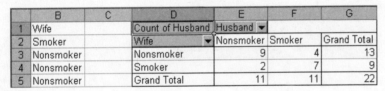

	B	C	D	E	F	G
1	Wife		Count of Husband	Husband ▼		
2	Smoker		Wife ▼	Nonsmoker	Smoker	Grand Total
3	Nonsmoker		Nonsmoker	9	4	13
4	Nonsmoker		Smoker	2	7	9
5	Nonsmoker		Grand Total	11	11	22

Using Formulas to Calculate Expected Cell Frequencies

You need the observed row and column frequencies to calculate the expected cell frequencies. For convenient access to these observed frequencies, let's place an expected frequencies table a couple of rows below the observed frequencies table.

1. Activate cell **D7** and key in the label **Expected Frequencies**. Also key in the row and column labels in the same way that they appear in the observed frequencies table at the top of the worksheet. You will use this table to display your work.

	B	C	D	E	F	G
1	Wife		Count of Husband	Husband ▾		
2	Smoker		Wife ▾	Nonsmoker	Smoker	Grand Total
3	Nonsmoker		Nonsmoker	9	4	13
4	Nonsmoker		Smoker	2	7	9
5	Nonsmoker		Grand Total	11	11	22
6	Nonsmoker					
7	Smoker		Expected Frequencies			
8	Nonsmoker			Husband		
9	Smoker		Wife	Nonsmoker	Smoker	
10	Smoker		Nonsmoker			
11	Nonsmoker		Smoker			

2. The calculation formula for the expected cell frequencies is

$$E_j = \frac{(\text{Column Grand Total})(\text{Row Grand Total})}{\text{Overall Grand Total}}$$

where the totals refer to grand totals displayed in the observed frequencies table. Each expected cell frequency is found by multiplying its column grand total by its row grand total and then dividing by the overall grand total (22).

For example, the expected frequency for the Wife Nonsmoker/Husband Nonsmoker cell would be computed as follows:

$$E = \frac{(11)(13)}{22} = 6.5$$

For efficiency and accuracy, I recommend clicking in cells instead of keying in the information. So, rather than using numerical values, we will be using cell addresses. The formula you will enter in the Wife Nonsmoker/Husband Nonsmoker cell is **=E5*G3/G5**. A similar formula will be used to calculate every expected cell frequency. I will give detailed instructions on how to calculate the expected frequency for only one cell. Then, I will provide the formulas for all the cells so that you can check your understanding of the procedure.

To begin, activate the Wife Nonsmoker/Husband Nonsmoker expected frequency cell (**E10**) by clicking in it.

3. Key in =.

Don't enter the period that follows the equal sign at the end of the sentence.

---

4. Click in cell **E5**.

5. Key in *.

6. Click in cell **G3**.

7. Key in /.

8. Click in cell **G5**.

9. Press [**Enter**]. Cell E10 should now contain an expected frequency of 6.5.

10. Repeat steps 2-9 to calculate the expected frequency for each of the remaining cells. To guide you in entering the information, the formulas are displayed below.

Expected Frequencies		
	Husband	
Wife	Nonsmoker	Smoker
Nonsmoker	=E5*G3/G5	=F5*G3/G5
Smoker	=E5*G4/G5	=F5*G4/G5

If the formulas have been entered accurately, the resulting numerical values will be the same as those displayed below.

	A	B	C	D	E	F	G
1	Husband	Wife		Count of Husband	Husband ▾		
2	Smoker	Smoker		Wife ▾	Nonsmoker	Smoker	Grand Total
3	Smoker	Nonsmoker		Nonsmoker	9	4	13
4	Nonsmoker	Nonsmoker		Smoker	2	7	9
5	Smoker	Nonsmoker		Grand Total	11	11	22
6	Nonsmoker	Nonsmoker					
7	Smoker	Smoker		Expected Frequencies			
8	Nonsmoker	Nonsmoker			Husband		
9	Nonsmoker	Smoker		Wife	Nonsmoker	Smoker	
10	Smoker	Smoker		Nonsmoker	6.5	6.5	
11	Nonsmoker	Nonsmoker		Smoker	4.5	4.5	

In the remaining sections of this chapter, I explain how to use Excel's functions to find the critical value of chi-square for a specified alpha, to find the P-value associated with the observed value of chi-square, and to calculate observed chi-square. The results of the calculations will be placed in the lower part of the worksheet.

Using Functions for a Chi-Square Test of Independence

Now that you have observed and expected frequencies displayed in a worksheet, you can utilize Excel's functions to finish the chi-square calculations for you.

1. Begin by keying in labels for **Alpha, df, Critical chi-square, P-value,** and **Observed chi-square** in column D as shown below.

	A	B	C	D	E	F	G
1	Husband	Wife		Count of Husband	Husband ▼		
2	Smoker	Smoker		Wife ▼	Nonsmoker	Smoker	Grand Total
3	Smoker	Nonsmoker		Nonsmoker	9	4	13
4	Nonsmoker	Nonsmoker		Smoker	2	7	9
5	Smoker	Nonsmoker		Grand Total	11	11	22
6	Nonsmoker	Nonsmoker					
7	Smoker	Smoker		Expected Frequencies			
8	Nonsmoker	Nonsmoker			Husband		
9	Nonsmoker	Smoker		Wife	Nonsmoker	Smoker	
10	Smoker	Smoker		Nonsmoker	6.5	6.5	
11	Nonsmoker	Nonsmoker		Smoker	4.5	4.5	
12	Smoker	Smoker					
13	Smoker	Nonsmoker		Alpha			
14	Nonsmoker	Nonsmoker		df			
15	Smoker	Smoker		Critical chi-square			
16	Smoker	Smoker		P-value			
17	Nonsmoker	Nonsmoker		Observed chi-square			

2. **Alpha**. Let's set alpha equal to .05 for this analysis. Activate cell **E13** and key in **.05**.

13	Smoker	Nonsmoker	Alpha	0.05
14	Nonsmoker	Nonsmoker	df	
15	Smoker	Smoker	Critical chi-square	
16	Smoker	Smoker	P-value	
17	Nonsmoker	Nonsmoker	Observed chi-square	

3. **df**. The formula for degrees of freedom (df) for the chi-square test of independence is $df = (r - 1)(c - 1)$ where r refers to the number of rows in the cross tabulation and c refers to the number of columns. Because this cross tabulation has 2 rows and 2 columns, $df = (2 - 1)(2 - 1) = 1$. Activate cell **E14** and enter **1**.

13	Smoker	Nonsmoker	Alpha	0.05
14	Nonsmoker	Nonsmoker	df	1
15	Smoker	Smoker	Critical chi-square	
16	Smoker	Smoker	P-value	
17	Nonsmoker	Nonsmoker	Observed chi-square	

4. **Critical chi-square**. You will use Excel's CHIINV function to find the critical chi-square value. Activate cell **E15**. Click **Insert** in the Menu Bar and select **Function**.

5. In the Insert Function dialog box, select the **Statistical** category, select the **CHIINV** function, and click **OK**.

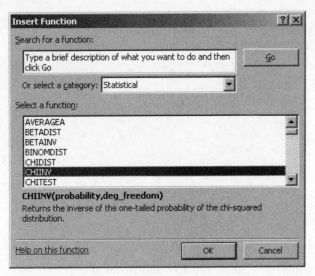

6. **Probability**. Click in the **Probability** window of the CHIINV dialog box and enter the value of alpha. To enter this value, you can either click in cell **E13** where the value of alpha has been entered, or you can manually enter **.05**. My preference is to use a cell address, which is what I have shown below.

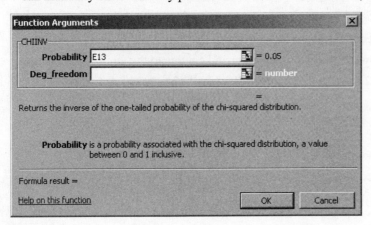

7. **Deg_freedom**. Click in the **Deg_freedom** window and enter the df value. To enter this value, you can either click on cell **E14** where df has been entered in the worksheet, or you can key in **1**.

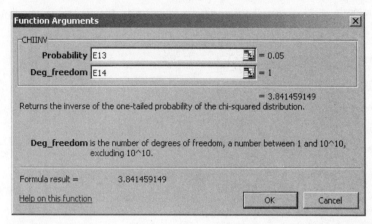

8. Click **OK**. The CHIINV function returns a critical chi-square value of 3.8415 and places it in cell E15 of the worksheet.

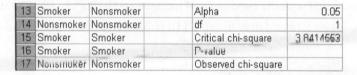

13	Smoker	Nonsmoker		Alpha	0.05
14	Nonsmoker	Nonsmoker		df	1
15	Smoker	Smoker		Critical chi-square	3.8414553
16	Smoker	Smoker		P-value	
17	Nonsmoker	Nonsmoker		Observed chi-square	

9. **P-value**. The P-value is the probability associated with the observed (or calculated) value of chi-square. You will use Excel's CHITEST function to find the P-value. Activate cell **E16**. Click **Insert** in the Menu Bar and select **Function**.

10. In the Insert Function dialog box, select the **Statistical** category, select the **CHITEST** function, and click **OK**.

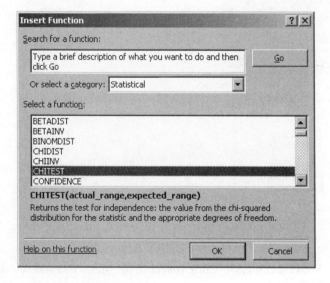

11. **Actual range** refers to the range associated with the observed cell frequencies. For this example, that range is E3:F4. Note that grand totals are not included in the range. Click in the **Actual_range** window of the CHITEST dialog box. Then click in cell **E3** and drag to cell **F4**. If you prefer, you can manually enter **E3:F4**.

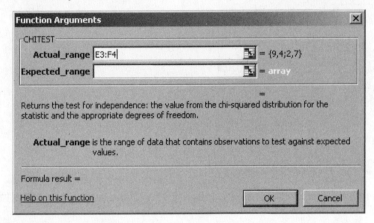

12. **Expected range** refers to the range associated with the expected cell frequencies, E10:F11. Click in the **Expected_range** window. Then click in cell **E10** and drag to cell **F11**. If you prefer, you can manually enter **E10:F11**.

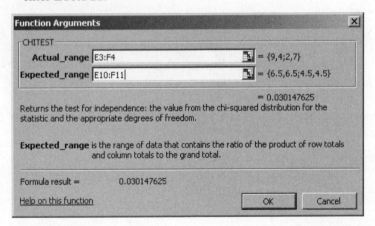

13. Click **OK**. The CHITEST function returns a P-value of .0301 and places it in cell E16 of the worksheet.

13	Smoker	Nonsmoker	Alpha	0.05
14	Nonsmoker	Nonsmoker	df	1
15	Smoker	Smoker	Critical chi-square	3.8414591
16	Smoker	Smoker	P-value	0.0301476
17	Nonsmoker	Nonsmoker	Observed chi-square	

14. **Observed chi-square**. You will use Excel's CHIINV function to find the observed chi-square value. You used the CHIINV function earlier in this example. Recall that CHIINV returns the chi-square associated with a specified P-value and df. Activate cell **E17**. Click **Insert** in the Menu Bar and select **Function**.

15. In the Paste Function dialog box, select the **Statistical** category, select the **CHIINV** function, and click **OK**.

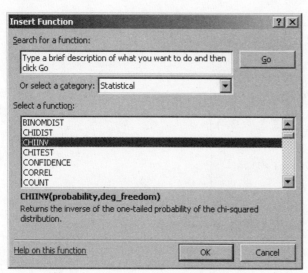

16. **Probability**. Click in the **Probability** window of the CHIINV dialog box and enter the P-value associated with obtained chi-square. To enter this value, you can either click in cell **E16** where the P-value appears in the worksheet or you can manually enter **E16** in the probability window.

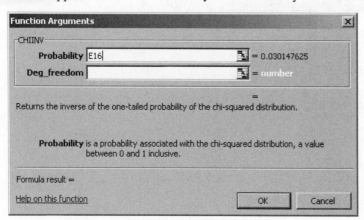

17. **Deg_freedom**. Click in the **Deg_freedom** window and enter the df value. To enter this value, you can click in cell **E14** where df has been entered in the worksheet, you can manually enter **E14**, or you can key in **1** (the numerical value of df).

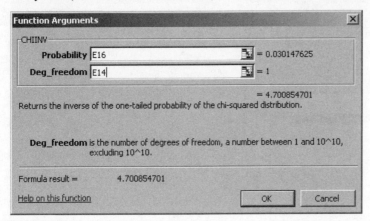

18.   Click **OK**. An observed chi-square value of 4.7009 is returned and placed in cell E17 of the worksheet.

This completes the chi-square test of independence. The worksheet with all entries is shown below.

	A	B	C	D	E	F	G
1	Husband	Wife		Count of Husband	Husband ▾		
2	Smoker	Smoker		Wife ▾	Nonsmoker	Smoker	Grand Total
3	Smoker	Nonsmoker		Nonsmoker	9	4	13
4	Nonsmoker	Nonsmoker		Smoker	2	7	9
5	Smoker	Nonsmoker		Grand Total	11	11	22
6	Nonsmoker	Nonsmoker					
7	Smoker	Smoker		Expected Frequencies			
8	Nonsmoker	Nonsmoker			Husband		
9	Nonsmoker	Smoker		Wife	Nonsmoker	Smoker	
10	Smoker	Smoker		Nonsmoker	6.5	6.5	
11	Nonsmoker	Nonsmoker		Smoker	4.5	4.5	
12	Smoker	Smoker					
13	Smoker	Nonsmoker		Alpha	0.05		
14	Nonsmoker	Nonsmoker		df	1		
15	Smoker	Smoker		Critical chi-square	3.8414591		
16	Smoker	Smoker		P-value	0.0301476		
17	Nonsmoker	Nonsmoker		Observed chi-square	4.7008547		
18	Nonsmoker	Nonsmoker					
19	Smoker	Smoker					
20	Smoker	Nonsmoker					
21	Nonsmoker	Nonsmoker					
22	Nonsmoker	Nonsmoker					
23	Nonsmoker	Smoker					

Chapter 13

Random Samples

Random Selection Using the Random Number Generation Tool

Researchers often need to select a random sample of people to participate in a research project. Excel's Random Number Generation tool provides an easy way to do this. The Random Number Generation tool samples with replacement, so it is possible that the same participant could be selected more than once.

Sample Research Problem

A customer service department wanted to survey a random sample of five customers who had phoned for assistance during the past week. Each of the 20 customers who had phoned was assigned a customer number from 1 to 20. The probability of selecting any one of these customers was 1/20 or .05.

1. Enter the labels, customer numbers, and the probabilities in an Excel worksheet as shown below.

	A	B	C
1	Customer	Probability	Sample
2	1	0.05	
3	2	0.05	
4	3	0.05	
5	4	0.05	
6	5	0.05	
7	6	0.05	
8	7	0.05	
9	8	0.05	
10	9	0.05	
11	10	0.05	
12	11	0.05	
13	12	0.05	
14	13	0.05	
15	14	0.05	
16	15	0.05	
17	16	0.05	
18	17	0.05	
19	18	0.05	
20	19	0.05	
21	20	0.05	

2. Click **Tools** in the Menu Bar and select **Data Analysis**.

If Data Analysis does not appear as a choice in the Tools menu, you will need to load the Microsoft Excel ToolPak add-in. Follow the procedure on page 7.

3. Select **Random Number Generation** in the Data Analysis dialog box. Click **OK**.

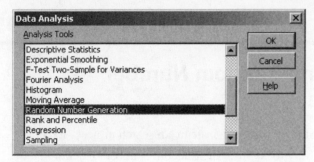

4. Complete the Random Number Generation dialog box as shown below. Click **OK**.

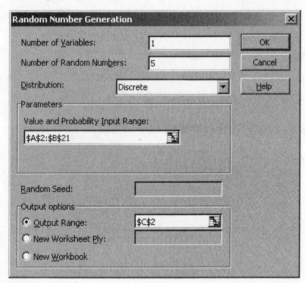

5. The random sample shown below contains a repetition of customer number 12. If repetitions occur in your sample, repeat the procedure to generate more numbers. Click **Tools** and select **Data Analysis**.

	A	B	C
1	Customer	Probability	Sample
2	1	0.05	1
3	2	0.05	6
4	3	0.05	12
5	4	0.05	8
6	5	0.05	12

6. Select **Random Number Generation** in the Data Analysis dialog box. Click **OK**.

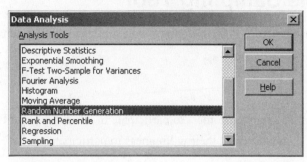

7. Complete the Random Number Generation dialog box as shown below. Be sure to use cell C7 for the output range. Click **OK**.

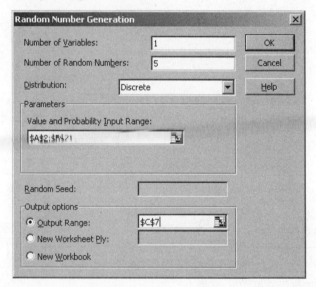

The list below now enables you to select five different customers for the survey project. Those customers are numbers 1, 6, 12, 8, and 4. If any of these customers decline to participate, you can continue with customer number 9.

	A	B	C
1	Customer	Probability	Sample
2	1	0.05	1
3	2	0.05	6
4	3	0.05	12
5	4	0.05	8
6	5	0.05	12
7	6	0.05	8
8	7	0.05	1
9	8	0.05	4
10	9	0.05	9
11	10	0.05	9

Random Selection Using the Sampling Tool

Samples can be selected randomly using Excel's Sampling tool.   The Sampling tool, similar to the Random Number Generation tool, samples with replacement.

Sample Research Problem

A market researcher wanted to select a random sample of eight customers to participate in a focus group. The researcher was selecting from a pool of 15 customers who indicated their willingness to participate.

1.   Enter the label **Customer** and the numbers 1 through 15 as shown below.

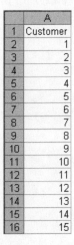

	A
1	Customer
2	1
3	2
4	3
5	4
6	5
7	6
8	7
9	8
10	9
11	10
12	11
13	12
14	13
15	14
16	15

2.   Click **Tools** in the Menu Bar and select **Data Analysis**.

If Data Analysis does not appear as a choice in the Tools menu, you will need to load the Microsoft Excel ToolPak add-in. Follow the procedure on page 7.

3.   Select **Sampling** in the Data Analysis dialog box. Click **OK**.

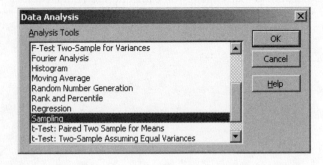

4. Complete the Sampling dialog box as shown below. Click **OK**.

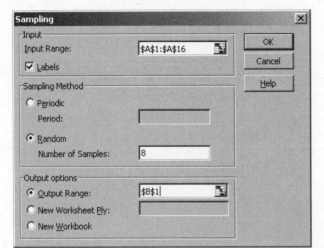

Because these numbers were generated randomly, it is not likely that your output will be exactly the same as mine. There were no repetitions in my output. The focus group will include customers who were assigned the numbers 1, 3, 4, 5, 6, 7, 11, and 15. If there were repetitions in your output, you can repeat the procedure until you have a total of eight different customers.

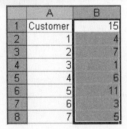

Random Selection Using the RANDBETWEEN Function

The RANDBETWEEN function returns a randomly selected number between the numbers specified by the researcher. This function is especially useful if the researcher is selecting from a very large range of numbers.

Sample Research Problem

An energy conservation firm in a large city wanted to find out if home buyers took energy-saving features into consideration when they bought their house. During the last year, the total number of home sales in the city was 2,000. The firm wanted to randomly select 20 of these home buyers to receive a free energy audit and to participate in a personal interview.

1. We will start by selecting one home buyer. Click in cell **A1** where the buyer's number will be placed.

The RANDBETWEEN function is in the ToolPak add-in. The function will return the #NAME? error if the Analysis ToolPak is not available. Instructions for loading the Microsoft Excel ToolPak add-in are given on page 7.

---

2. Click **Insert** in the Menu Bar and select **Function**.

3. Select the **All** category. Then scroll down to select the **RANDBETWEEN** function. Click **OK**.

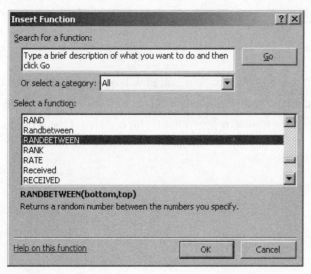

4. Type **1** in the Top window. Type **2000** in the Bottom window. Click **OK**.

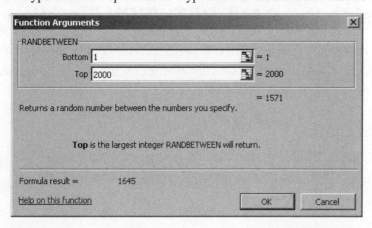

5. The function returns 258. Because the number was generated randomly, it is not likely that your output will be the same.

6. The firm wants a total of 20 randomly selected home buyers. Copy the function in cell **A1** to cells A2 through A20.

	A
1	1853
2	562
3	960
4	1623
5	1758
6	1750
7	1896
8	985
9	419
10	566
11	1049
12	1529
13	1951
14	543
15	1025
16	561
17	1404
18	1790
19	1466
20	1607

Note that the number in cell A1 changed. This is because Excel will return new random numbers every time the worksheet is calculated. The RANDBETWEEN function samples with replacement. So, it is possible that your list of 20 numbers will have repetitions. To make sure that you have 20 different numbers, you may want to generate a sample of 25 to 30.

Index